D0438284

THE KILLER STRAIN

Also by Marilyn W. Thompson

Feeding the Beast
Ol' Strom

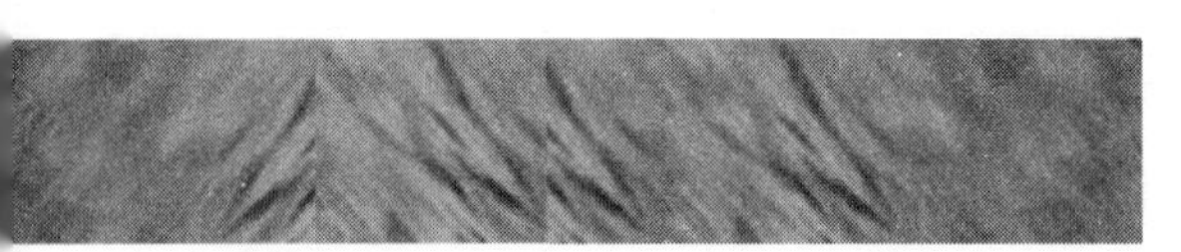

Marilyn W. Thompson

THE KILLER STRAIN

Anthrax and a Government Exposed

HarperCollins*Publishers*

THE KILLER STRAIN. Copyright © 2003 by Marilyn W. Thompson. All rights reserved. Printed in the United States of America. No part of this book may be used or reproduced in any manner whatsoever without written permission except in the case of brief quotations embodied in critical articles and reviews. For information, address HarperCollins Publishers Inc., 10 East 53rd Street, New York, NY 10022.

HarperCollins books may be purchased for educational, business, or sales promotional use. For information, please write: Special Markets Department, HarperCollins Publishers Inc., 10 East 53rd Street, New York, NY 10022.

FIRST EDITION

Designed by Sarah Maya Gubkin

Printed on acid-free paper

Library of Congress Cataloging-in-Publication Data

Thompson, Marilyn W.
The killer strain : anthrax and a government exposed / Marilyn W. Thompson.—1st ed.
p. cm.
ISBN 0-06-052278-X
1. Anthrax—United States. 2. Bioterrorism—United States. I. Title.

RA644.A6T48 2003
362.1'96956—dc21

2002191933

03 04 05 06 07 ❖/RRD 10 9 8 7 6 5 4 3 2 1

To my parents, Hugh and Jessie Walser

Contents

The Eclipse

CHAPTER 1

In the Field

October 19, 2001
STAFFORD, VIRGINIA

By the time Leroy Richmond awoke, the lethal spores had settled into his lungs, but of course he did not know it. He felt hot and achy, and wondered if he might be coming down with the flu. It was before dawn on Friday, a regular workday at the U.S. Postal Service, and Richmond did not indulge thoughts of staying home. His wife, Susan, often complained that he was married to the job—a "workaholic"—but he looked forward to each day of handling express mail at the cavernous Brentwood Mail Processing and Distribution Center in Washington, D.C.

Richmond slowly rolled out of bed and washed and dressed as usual, trying to ignore the erratic fever he had been battling for several days. He would feel bad and then suddenly better, a phenomenon known in the medical literature as an "eclipse." He had been treating his symptoms with common aspirin, a laughable remedy given the virulent nature of the bacteria infecting him, like confronting an attacking tiger with a pellet gun. The aspirin made him feel better, but the relief was perilously deceptive. Microscopic rod-shaped germs arrayed in long, narrow chains incubated in the warm recesses of his chest, mustering for a stealthy assault. Within hours, they would send two toxins surging through Richmond's bloodstream, poisons that could render powerless the most potent treatments. His lungs would bleed and swell with germ-clouded liquid. The pressure would threaten his heart, and he would drift in and out of consciousness, breathing weakly through a respirator.[1] Statistically, he had a slim chance of beating the pathogen unleashed inside his body.

The clock read 2:50 A.M. as Richmond tiptoed through the dark, past polished tables filled with photographs of family, some of whom were still sleeping in the rooms behind him. His wife kept their two-story house immaculate, the white overstuffed furniture in the living room spotless, the dining-room table set as if for company, with cloth napkins tucked into crystal glasses. They had lived in the spacious home for seven years, a testament to their upward striving. Richmond, a tall, slender man born in Newport News, Virginia, had worked for USPS for thirty-two years, mostly at Brentwood and its predecessor on North Capitol Street. Brentwood was so full of old-timers that it felt like a second home. Everyone there called him Rich, never Leroy. He had met the feisty Susan there, working the line.

It was not an instant attraction. One day, a supervisor sent her over to help him manually sort mail. To the industrious Rich, all she seemed to do was complain. She was tired. She didn't feel well. Before long, he caught her catnapping.

He asked his boss not to send her over again. The next day, there she was, grumbling, napping, disappearing for long breaks. Rich asked her to speed up, and she shot him a cutting look and barked, "You're not my supervisor!" Rich went back to the boss and suggested that he fire her.

Rich ran into Susan sometime later at a club, dancing, turning on the charm. He was mesmerized. He couldn't get her off his mind, her broad hips and beautiful braids. A fiery courtship began, and they were married in less than a year.

Now that their youngest child, Quentin, was seven, they found working alternate Brentwood shifts the best way of managing their hectic lives. The routine was taxing. Susan had come home from that night's shift and crawled into bed after 1 A.M., just before Rich's day began. She noticed him feebly dressing for work, coughing, looking gaunt and worn after several days of inexplicable tiredness. Not mincing words, she called out:

"You look like a crack addict! Where are you going?"[2]

"Going to work." He sighed.

She scowled as he downed more aspirin and finished preparing to leave.

"Me?" she would say later, standing defiantly with hands on hips. "I take some medication and roll over. He'll have a hundred-and-two-degree fever and go to work!"

True, it had been years since Rich had called in sick. He was more likely to volunteer for overtime—anything to keep the money rolling in.

Anyway, there seemed to be little use asking for time off from Brentwood's hard-line management. Unspoken tension divided the center between the almost exclusively African-American workforce, stationed behind chugging machines and conveyor belts, and the many white supervisors patrolling the production lines. The U.S. Postal Service (USPS) touted its minority hiring as a sign of progressive management, but among some of the workers, the atmosphere so harkened back to the plantation days of the Old South that they derisively referred to Brentwood's floor as "the field."

Rich tried not to think in such stark racial terms. He only knew that the last time he had asked for time off, to attend his now-grown daughter's school play, his supervisor had scoffed at him. He had never asked for a favor again.

Besides, he took pride in postal work and knew he was good at it. Name a street address, a federal building or embassy anywhere in Washington, and he could rattle off the zip code from memory, the result of untold hours of training. Years ago, he had walked the streets of suburban Washington to stuff letters into mailboxes, so he knew firsthand that Americans took seriously the credo that "neither snow, nor rain" would stop the U.S. mail. There was something exhilarating about leaving a mailbox full of letters and colorful postcards, third-class catalogs and cumbersome junk mail, then watching expectant old ladies and children rush out the door to retrieve their surprises as he walked away. Sometimes, they peeked from behind curtains mysteriously, trying to speed the process. He was Santa Claus in a blue uniform.

That morning, sweating and breathing hard, his head pounding, Rich climbed into his blue 1980 Sierra that had seen too many miles and pulled resolutely out of the asphalt driveway, setting off on the fifty-three-mile commute into Washington. He and Susan had been lured to Stafford by affordable housing that made it possible for a working couple to live like the rich, with enough bedrooms to accommodate visiting relatives and big lawns with azaleas and backyard swing sets. He made a left out of the tight cul-de-sac, winding through the neighborhood of cookie-cutter designs and two- and three-car garages. A sharp right took him onto Highway 610, the main drag through Stafford, past the gas station where he usually stopped for coffee. After zooming through a chain of traffic lights along the darkened commercial strip, past empty strip malls and sprawling discount stores, he finally exited onto Interstate 95 toward Washington.

At this time every morning, Rich realized the benefit of his ungodly hours. His schedule allowed him to avoid one of the worst commuter nightmares in America, a bumper-to-bumper backup of Washington-bound bureaucrats and federal worker bees that started as early as 5 A.M. and seemed never to stop. The roads were so predictably jammed during rush hours that many drivers took the risk of picking up strangers at bus stops so they could zip into the High Occupancy Vehicle lanes, which required three or more passengers to enter. Rich did not have to bother with such nonsense. He could drive undeterred through the lush Virginia countryside, where wildlife often ambled wide-eyed across the highway, and past the sprawling Potomac Mills Outlet Mall, in daytime hours a dreaded bottleneck of tourists and bargain hunters. He could even breeze past the notorious congestion center along I-95 known as "the Mixing Bowl," which blended the crisscrossing highways of Arlington and Alexandria.

Rich zipped through the interchange and passed the Pentagon. Since September 11, one side of the building stood in rubble where terrorists had rammed a hijacked jetliner into it, and workers sifted through the charred remains. Such distractions were best ignored. Rich's drive time, sixty-five minutes, was unfailingly predictable if he stayed on course. He would leave I-95, exit onto the Southeast

Freeway to New York Avenue, wind his way back to Brentwood Street NE, and pull into the postal center's sprawling parking lot at 3:55 A.M.

The radio in the Sierra had broken ages ago, but Rich didn't miss it. He was not a news junkie, addicted to traffic reports or the drone of twenty-four-hour newscasts. However, he couldn't avoid the news on September 11. On the day of the attacks, he happened to be walking through Brentwood's cafeteria on an errand about the time the second hijacked jetliner crashed into the World Trade Center. He saw it all on the cafeteria television, the only one on the floor, and was sickened. But Rich had composed himself and gone to work, and the production line never stopped, even when Brentwood workers noticed the black smoke from the blazing Pentagon clouding the sky over Washington.

In the tumultuous weeks after 9/11, most of Brentwood's workers turned to God for solace. They were a deeply religious lot, and Rich often joined a small devotional group that met each morning around 4:15. The leader was the devout Joseph Curseen from southern Maryland, who had worked Brentwood's night shift for fifteen years. Curseen, who carried a distinctive red Bible, usually chose the daily Scripture reading. He would read it aloud, then ask, "What does it mean to you?" The group would have a lively discussion, while others on the floor used the work break to snack, gossip, or catnap. Rich found the prayer group comforting.[3]

Somehow, Rich remained oblivious to a chain of events that had begun two weeks earlier in Florida, a medical mystery that had gradually emerged as a second sinister act of terrorism. A sixty-three-year-old tabloid photo editor had died, and a seventy-three-year-old mailroom worker from the same Boca Raton media company lay critically ill in a Miami hospital. Authorities were stunned; the illnesses had been linked to a bacteria known as *Bacillus anthracis,* a germ that most commonly infects goats and sheep and rarely afflicts humans, and then usually only those who have close contact with animals or their hides. The Florida men worked together in an office building, the headquarters of the supermarket tabloids the *Sun* and the *National Enquirer.* They had both contracted inhalation anthrax disease, an extremely rare occurrence in the United States. Despite the best damage-control efforts of civil authorities, the events were the realization of a long-standing fear, an act of bioterrorism on American soil.

Anthrax, its name taken from the Greek word for "coal" has a fascinating history reaching back five thousand years. The fifth and sixth plagues of Egypt, described in the book of Exodus, are believed to have been anthrax, as is the devastating Black Bane that swept through Europe in medieval times.

In cultures, *B. anthracis* has a distinctive rod shape and often clumps together in long rods resembling a bamboo shoot, surrounded by a protective coating. It produces spores that can live dormant in the soil for years, perhaps decades, waiting for an unsuspecting host. Amazingly hearty, the spores can withstand heat and cold, even disinfectants and blasts of radiation. They begin to vegetate and multiply when they enter the nurturing environment of a human or animal host.

The diseases caused by *B. anthracis* come in three varieties, all known in the vernacular simply as anthrax. The most common type, which accounts for ninety-five percent of all cases worldwide, is the cutaneous (skin-based) form, characterized by lesions that develop black scabs at the point of contact. Skin anthrax is rarely fatal and can be effectively treated with antibiotics. In the United States, 224 cases of skin anthrax were documented between 1944 and 1994, with another single case reported in 2000. In the modern world's largest recorded outbreak, about ten-thousand cases surfaced in Zimbabwe between 1979 and 1985, an epidemic later believed to be the result of an experiment in germ warfare run by the Rhodesian military.

The gastrointestinal form of anthrax, linked to eating raw or undercooked anthrax-infected meat, causes violent vomiting and diarrhea. About half of its victims die from it. It is seen most commonly in Africa and Asia, in villages with poor meat handling.

The inhalation form was so prevalent among millworkers and tanners in nineteenth-century England that the British named it woolsorter's disease. Like its counterparts, it seldom appeared in modern times. In the United States, only eighteen cases were reported in the twentieth century, most of them in groups at high risk of exposure, such as textile workers and cattlemen. Doctors, influenced by animal studies, believed humans could not get the infection without inhaling at least several thousand spores. Each spore is no larger than one-and-a-half to two microns in size, a fraction of the width of a human hair.

Once contracted, though, this form of anthrax is nearly always fatal. Spores settle into the lymph nodes around the lungs; in this ideal growth climate, the dormant bacteria revive and rapidly multiply.

Anthrax's powers make it an attractive option for armies intent on creating weapons of mass destruction. Finding *B. anthracis* is as easy as locating the corpse of a disease-stricken animal in a field and digging up spores with the contaminated soil around it. Once spores are found, any competent microbiologist can cultivate them.

Many of the eighty-nine known strains of anthrax also may be purchased by credentialed researchers through a network of germ mail-order houses around the world. The American Type Culture Collection in suburban Washington came under fire in the late 1980s for shipping several varieties of anthrax and other pathogens to Iraq, a misstep that gave a sworn U.S. enemy seed material for a potential bioweapons arsenal. Until the late 1990s, when a domestic anthrax scare motivated the U.S. Congress to pass a law instituting inventory controls, some researchers freely swapped samples of bacteria used in their experiments. Even after the law was passed, some scientists suggested it was common to exchange germs among friends.

The United States military experimented with anthrax in its offensive bioweapons program, but President Richard Nixon halted the initiative in 1969. In 1972, the United States joined an alliance of 144 nations agreeing under the Biological and Toxins Weapons Convention Treaty that they would not produce chemical or biological weapons for offensive purposes. At U.S. military facilities involved in bioweapons production, stockpiles of bacterial agents were destroyed, including 220 pounds of weapons-quality anthrax.

Though the threat subsided for a time, by the 1990s military leaders in the United States became convinced that their enemies were still intent on producing and stockpiling biological weapons, and that this would one day surface as a major threat to national security. In November 2001, at a biological weapons conference in Geneva, the administration of President George W. Bush boldly voiced its belief that Iraq and five other nations—North Korea, Libya, Syria, Iran, and Sudan—were actively pursuing germ warfare programs. The declaration was part of an effort to build an international coalition against Iraq's Saddam Hussein, who had kicked out United Nations weapons inspectors in 1997 and was under increasing U.S. pressure, including the threat of war, to allow them to return. U.S. urgency had been heightened by the catastrophe of 9/11. Materials seized from terrorist training camps in Afghanistan suggested that the international network led by Osama bin Laden had the desire and the capability to secure anthrax and use it as a weapon of mass destruction. Iraq was considered bin Laden's most likely source.

Military and counterterrorism experts in the United States considered anthrax ideal for a bioterror assault: it could be inexpensively grown from a few spores into a truckload, and according to a World Health Organization estimate, the release of fifty kilos of anthrax in a city of 500,000 would cause 125,000 infections and 95,000 deaths in just three days.

There was ample proof of anthrax's lethality. In 1979, anthrax was accidentally spewed into the air at a secret Soviet military installation in Sverdlovsk, killing at least sixty-eight people. The Soviets initially denied that the deaths were linked to weapons production, suggesting instead that they had been caused by tainted meat. It was not until a 1992 summit meeting that Soviet President Boris Yeltsin acknowledged that the accident had occurred during the development of anthrax weapons.

Until the Florida cases surfaced, however, most Americans dismissed attention to anthrax as mere hysteria, preached by a handful of self-appointed bioterror experts. The nation's most serious case in 2000, for example, had nothing to do with terrorism; four North Dakota cows diagnosed with anthrax caused a quarantine of thirty-two cattle farms. Only one human was infected, a sixty-seven-year-old North Dakota rancher who contracted skin anthrax while disposing of the cattle carcasses.[4]

The Florida episodes sent seasoned doctors back to their microbiology textbooks. The first victim, photo editor Bob Stevens, walked into a Boca Raton emergency room on his own on October 2, 2001, went into a coma, and quickly died. The rarely seen bacteria clouding his spinal fluid confounded doctors.

The case sent a collection of federal agencies into Florida, among them the Federal Bureau of Investigation (FBI), the Centers for Disease Control (CDC), and the Environmental Protection Agency (EPA). Their activities were closely tracked by the Bush White House and monitored through daily phone conferences coordinated by the National Security Council.

From the beginning, there was an effort to avoid any action that would incite panic. When health teams began to suspect that a piece of mail might be involved, postal officials insisted that nothing should be done to interrupt the delivery of the U.S. mail. Public statements from Bush and his cabinet urged calm and caution, while health officials quietly mobilized for what they feared could be a larger-scale bioterrorist attack.

Secretary of Health and Human Services Tommy G. Thompson, the nation's top health official, went on television to speculate that Stevens's death could have been caused by something as innocuous as water from a mountain stream along a trail where Stevens had been hiking recently in North Carolina. Disease hunters from the CDC combed the trail and its surroundings looking for any possible anthrax source—a dead cow, a tourist shop marketing Indian drums made from animal hides. They came up with nothing.

In Florida, another team sifted through Stevens's Lantana home, canvassed his regular biking trail, and surveyed stores where he liked to buy exotic imported spices and Indian foods. They scouted Florida hospitals for other possible cases.

This broad-based survey yielded a substantial clue, a second possible anthrax victim. A Miami doctor reported that he, too, had identified rod-shaped bacteria in a patient then hospitalized with pneumonia. Ernesto Blanco was a friend of Stevens's who worked in the same building, the Boca Raton headquarters of American Media Inc., publisher of the *Enquirer* and the *Sun*. Authorities sealed off the AMI building to search for evidence and walked through the contaminated building unprotected. Employees abandoned purses, lunch boxes, even pet goldfish, as they evacuated, a nightmarish scare that seemed ripped from their own tabloid headlines.

The feds carted off Stevens's office computer keyboard and mail slot, chunks of carpeting, clothing, and equipment for testing. Yet no one could imagine the intentional targeting of Stevens, a mild-mannered transplanted Brit who spent most of his time behind a desk, or Blanco, a genial grandfather working beyond retirement age to deliver piles of office mail. As AMI employees lined up for antibiotics, one speculated to Palm Beach County Health Department Director Dr. Jean Marie Malecki that the building might had been singled out for vengeance because the tabloids had interviewed a supposed concubine of terrorist leader bin Laden who complained about the inadequacies of their sex life.

As facts emerged, federal authorities seemed most alarmed by employees' memories of a suspicious piece of mail delivered a few days earlier. Written as a fan letter to singer Jennifer Lopez and bearing a Star of David, it had emitted a puff of white powder when it was opened. Some remembered that Stevens held the letter close to his face to examine it. A second suspicious letter also had generated attention, but both had been discarded and incinerated by the time Stevens died.

These recollections finally caused authorities to voice the fear that the case was bioterrorism, a new round of murder in the wake of September 11. They downplayed the possibility that the same group of terrorists was responsible, although its ties to South Florida were extensive.

Suddenly mail in other places became more sinister. When a New York doctor reported that he had been treating an assistant to NBC news anchor Tom Brokaw for a lesion that could be cutaneous anthrax, another team of federal specialists sped to New York to investigate. An editorial aide at the *New York Post* reported a classic black-scabbed lesion. A seven-month-old baby brought into his mother's office at ABC-TV had developed a red blister on his arm, and then fell deathly ill days later when anthrax toxins surged through his tiny system.

A pattern began evolving. Almost all of the incidents involved letters delivered by regular mail through the postal service. They were written in a similar primitive script, and carried similar death threats. Conspicuously, they offered praise to Allah. The return address scrawled in the corner of two of them was to a nonexistent elementary school in New Jersey. Filled with powder, the letters were sealed shut and seemed unremarkable before recipients ripped them open and saw a whoosh of powder.

The implications of anthrax letters passing undetected through the postal system did not dawn on officials or on the teams of health experts working the anthrax cases on the ground. At the teeming Brentwood mail center, where three million pieces of mail were processed each day, workers like Rich remained unaware of possible workplace dangers, even during the "safety talks" that were part of every workday. With a lilting accent that suggested Caribbean heritage, the popular Rich had often been asked by his supervisors to read the daily lessons to a cluster of about fifteen employees. Several messages since 9/11 had asked employees to be alert for possible acts of terrorism, but the specifics of the Florida cases had not been publicized.

Brentwood had taken notice of the anthrax scare on October 15, when another letter was found at the Hart Senate Office Building on Capitol Hill. A female intern working for Senator Tom Daschle (D-S.D.), the Senate majority leader, had opened a crudely handwritten letter that appeared to be addressed by a New Jersey schoolchild, still untrained in cursive script.

When she opened the letter, the fine white powder packed into the letter dispersed around her like a fine perfume. She screamed, attracting concerned friends and curious staffers. Unbeknownst to those present, the powder spread through Daschle's two-story office and seeped into a neighboring office suite. The first few Capitol police officers who responded to the emergency call unwittingly made matters worse as they tramped through the office, exposing themselves in the process. The substance was so light, so infinitely mobile, that within minutes, it turned a federal office building into a hazardous waste site.

Bar codes on the prestamped envelope detailed the letter's path. Postmarked in Trenton, New Jersey, it had arrived at Brentwood in a packed mailbag late on October 11 or early on October 12. The bag's contents were sorted into bar-coded trays and moved by conveyer belt to a large tray-sorting machine. The letter was manually fed into Machine No. 17 at 7:10 A.M., then conveyed to Brentwood's government mail section. It arrived at the Senate mailroom around noon on October 12.

Sometime between 8 and 9:40 A.M., Machine No. 17 was opened for routine cleaning. The $500,000 piece of equipment was a vital cog in the process of channeling the daily mail, and it had to be tended with care. To clean it, workers sprayed compressed air into the machine and the surrounding conveyer belts.[5]

But for several critical days, the possible path of human exposure to the anthrax in the Daschle letter went unexplored by federal authorities. On October 18, Postmaster General John Potter and other officials used Brentwood for a media event to announce a $1 million reward for information leading to the capture of the Daschle letter's sender. The setting was chosen to allay mounting concerns about the safety of the mail.

Later that same night, working her usual shift, Susan had listened while plant manager Tim Haney told the workers not to be alarmed when they saw hazardous materials teams testing the machines for anthrax. Haney was wearing no protective clothing, so the threat of anthrax seemed distant.

Only the bold questioned Haney's comments.

"Machines?" Susan heard someone angrily call out. "What about people?" Another suggested that Brentwood should be shut down like the Hart Building, as a precaution until its safety could be proved.

Susan listened with disgust as the conversation turned to grave estimates of what it would cost to close down Brentwood and the layoffs that would surely follow. The workers quieted down and dutifully returned to their stations.

The atmosphere was much different on Capitol Hill. Health officials conducted a massive program of nasal swabbing, and doled out Ciprofloxacin, the recommended antibiotic, to Hill staffers and visitors. The FBI swept up the deadly letter and five other pieces of mail from the Daschle intern's desk and whisked them away, along with the young woman's tainted dress and dozens of other samples. All of it went for expert testing to the U.S. Army Medical Research Institute for Infectious Diseases (USAMRIID) at Fort Detrick, Maryland. There, the anthrax evidence that would be critical in building a criminal case was secluded in an undisclosed location, accessible only to a few approved hands.

As facts began to emerge about the evidence, however, there was considerable irony in Fort Detrick's guardianship of the specimens. Within a few days, government researchers, using DNA analysis, had identified the specific strain of anthrax used in the attacks, which turned out to be one used mostly by the U.S. military and its trusted contractors. Scientists capable of making the refined product probably came down to no more than fifty people. A few possible suspects emerged. Homes were searched, then yards and outbuildings; but the killer's trail had gone cold.

As these theories percolated in the first weeks of the anthrax scare, federal authorities seemed much more confident in their assumptions about the safety of the postal system. Doctors from the CDC concluded that anthrax would not escape from a sealed envelope, and recommended that the Brentwood workers not take "Cipro," the same drug being handed out like candy on Capitol Hill.

To mollify workers, postal authorities promised 86 million pairs of vinyl gloves—three pairs per employee—and four million face masks to protect against contaminants. Susan wondered how the white gloves would help, since a fine white powder was unlikely to show up on their surface. Leroy Richmond looked for the gloves and masks but couldn't find them; he said he was told supplies had run out.

On the day the Daschle letter passed without notice through Brentwood, Rich had followed his normal routine: in by 4 A.M. to work "the sweetest job any man could want." He loved working with overnight mail because it had real urgency. Deliver it by noon and you could be "saving someone's mortgage, someone's house." His work area was a cubicle, shielded somewhat from the cacophony of the open floor. After a few hours in the District, he boarded a bus for the express mail center at Baltimore-Washington Airport and worked there until 9:20 A.M., then headed back to Brentwood for an early lunch. He liked hopping between two facilities. It gave the job variety.

When a supervisor ordered him to clean up behind Machine No. 17, which Joe Curseen had shut off as he went off duty around 8:05

A.M., Rich wasn't hesitant. He considered no chore beneath him and always did as he was told. The area was crowded with boxes, which Rich piled high on a cart to haul away. He remembered feeling a rush of air when a cleaning man, wearing protective clothing, blasted the idle machine in front of him with pressurized air. It would take days before officials realized that this routine cleaning had spread anthrax throughout Brentwood.

Few alarms went off at Brentwood, even when news surfaced that a postal worker in the Trenton, New Jersey, area had developed an anthrax lesion. Some of the suspect letters—including the one to Daschle—bore a Trenton postmark, meaning they had gone through a distribution center in Hamilton, New Jersey, near the state capital. However, guided by the CDC, New Jersey officials reassured workers that they were safe.

Rich had not yet made the connection between this chain of events and his own flulike symptoms. On the morning of the nineteenth, he sank weakly into the driver's seat and followed his usual driving ritual. He pulled a rosary from his right pocket and began to pray. He had grown up a Baptist in a deeply religious family, one of ten children, eight boys and two girls, with two sets of twins. At one point he had studied Catholicism, and he took comfort in its rituals.

Some days on the commute, he prayed for his twin brother, who still lived back in Newport News and had respiratory problems. Some days, he prayed for his father, suffering from Alzheimer's disease. But on this October day, Rich drove distractedly through the wounded city and wondered where he would find the strength to work a full day. He fingered the rosary and prayed for world peace.

Rich would not know until many delirious weeks later that other postal workers were feeling similarly ill that Friday morning, both at Brentwood and Hamilton. Curseen, a healthy forty-seven-year-old, wondered if he was coming down with a cold. Thomas "Mo" Morris Jr., fifty-five, another longtime Brentwood worker and a fixture in his Suitland, Maryland, bowling league, felt a flulike fatigue. Though they all knew one another and chatted when their paths crossed, the

Brentwood workers had no reason at this point to commiserate about their assorted aches and pains.

Meanwhile, in New Jersey, postal worker Teresa Heller's skin lesion failed to respond to routine treatment. Letter sorter Norma Wallace had begun vomiting on the weekend, but struggled through work with fever, chills, and breathlessness before seeing a doctor. He asked her to come back on Monday if she didn't feel better.

She couldn't wait until Monday. With worsening chest pain, Wallace went to see her primary physician on Friday the nineteenth, the same day that Rich's symptoms worsened. The doctor had heard about the Hamilton plant's role in the mailings and immediately sent Wallace to the emergency room. Despite her labored breathing and a suspicious scab on the back of her neck, doctors were uncertain at first that the case was inhalation anthrax. It would take several days to be sure.

The malaise and erratic symptoms of the postal workers disguised the awful reality of a disease whose course reads like science fiction. Once inhaled, minute anthrax spores attach themselves to the lungs and lie dormant for one to six days. The victim has no reason to suspect a potentially fatal illness. Even when mild symptoms surface, they sometimes improve dramatically before the patient begins a rapid, downward plunge.

The onset of fever and flulike pain usually means that the often-deadly process has begun. After a few days, the dormant spores are gobbled up by macrophages ("big eaters"), scavenger cells that are a natural part of the human immune system and roam the body, usually to protect it from disease. Anthrax turns the function of the macrophage on its head. The cells unwittingly become the bacteria's hosts, offering a free ride into the body's lymph system. There, protected inside the lymph nodes, the spores multiply and produce toxins, which flood into the bloodstream. Inhalation anthrax produces "pleural effusions" or liquids tainted with bacteria, which submerge the lungs. The disease often causes the lungs and lymph nodes to bleed, the brain to swell, and the heart, surrounded by fluids, to falter. Death by anthrax is quick and excruciating.

Anthrax cannot be transmitted from person to person, so patients dying from it do not have to be ministered to through protective suits and gloves, like victims of the untreatable "hot viruses" Ebola and Marburg. But anthrax victims, both animal and human, become incubators for the production of more anthrax spores. For this reason, veterinarians recommend that the corpses of anthrax-infected animals found dead in the field be burned on the spot. Only cremation ensures that the germ is destroyed and does not revive years later.

In the African plains, where anthrax is a naturally occurring disease, the bacteria is powerful enough to make a stricken elephant collapse to its knees, roll to its side, and die after the briefest illness. Unless someone intervened, the infected postal workers faced the same fate.

Rich arrived at Brentwood right on time on October 19, but by then, he could barely find the energy to get out of the car and walk to the nurses' station. He was feeling worse by the minute, the fever surging up, then crashing. He slowly opened the door and planted a foot on the ground. Climbing out of the car, he grasped the steering wheel for balance.

Once inside, sweating profusely, he looked for a supervisor who could sign a pink slip, giving him permission to see the nurse. Finally, he spotted a manager standing before a small group of workers, reading one of the daily health alerts that had grown progressively more alarming over the past few days. He asked for a pink slip, but the distracted supervisor brushed him aside.

Rich knew the mandatory health discussion would take twenty minutes, perhaps more if tough questions surfaced, and he worried that he couldn't wait that long. He went upstairs to the management offices, found a sick slip himself, filled it out, and finally found a boss to scribble approval.

The nurse checked his vital signs and temperature, which had dipped back to normal. She pronounced that she could find nothing wrong with him.

"I feel bad. I feel weak. I've been coughing up phlegm," he told her.

The nurse advised him to leave work and go to see a doctor. Rich called Susan to tell her he was heading straight for their HMO, the Kaiser Permanente clinic in Woodbridge, Virginia.

By this point, the anthrax bacteria incubating in Rich's body were about to unleash their poisons. The anthrax was the Ames strain, a clue that would lead authorities to conclude some weeks later that the attack was not foreign but most likely the work of a homegrown terrorist. Someone had managed to secure the strain and was using it to send a wake-up call to the U.S. government. Unless treated quickly, Rich, a healthy man only a week before, would become a human sacrifice to the killer's twisted cause.

CHAPTER 2

Red Sky at Morning

June 2001
ANNAPOLIS, MARYLAND

On an early summer morning, as sailors embarked from city docks and swimmers plunged into the briny water for an annual race across the Chesapeake Bay, John W. Ezzell and about two hundred other scientists convened on a deserted college campus in the colonial seaport town of Annapolis for the Fourth International Anthrax Conference. All of the researchers specialized in the study of *B. anthracis,* a bacterium that had long intrigued scientists. In the late nineteenth century, Robert Koch had built the foundation for the field of microbiology with his discovery of the origins of anthrax disease and its links to specific bacteria. The French scientist Louis Pasteur had been so interested in the disease destroying flocks of sheep across the countryside that he set out to create a vaccine that would protect them, unveiling it in 1881 at the request of the French Ministry of Agriculture. Pasteur also had been the first to report that animals infected with *B. anthracis* should not be buried, since anthrax could be spread from carcasses.

From the local to the federal level, costly preparedness drills had become common in the United States, but laboratories, hospitals, and police and fire departments were still uniformly unprepared to respond to an assault with anthrax or other bioweapons of mass destruction. Again and again, they had shown that no facet of the system could adequately respond to a catastrophe that could wipe out thousands in a single devastating blow. To better the situation, private companies had developed equipment to improve detection of deadly agents. Government consultants and agencies pondered such awful questions as how best to transport thousands of human casualties out of a paralyzed city. Federal health agencies beefed up contact with local laboratories and hospitals so the system could spring into action if disaster struck. The challenge remained daunting. Just that summer, the federal response system prepared to stage yet another drill, an elaborate computer-simulated bioterror exercise code-named "Dark Winter" that would feature role-playing by politicians, bioterror authorities, even journalists. Former Senator Sam Nunn, a Georgia Democratic powerhouse and longtime chairman of the all-important Appropriations Committee, had been cast as the president of the United States, grappling with a terrorist release of smallpox. At the end of these sessions, thick reports would document system inadequacies, and preparedness champions on Capitol Hill would make empty promises for more funding to fix them. Year after year, the ritual was repeated at local, state, and federal levels.

In the world of Ezzell and other anthrax specialists, the terror threat had real-life implications. For decades, scientists in the field had commonly swapped sensitive information, even shipping anthrax samples to colleagues for research. Some members of the tight-knit scientific community thought little of traveling to meetings with vials of *B. anthracis* samples to share with trusted colleagues as a professional courtesy.

By the opening of the Annapolis conference, this trust had begun to erode. Security had tightened dramatically at laboratories around the world, and scientists who dared to smuggle anthrax samples into or out of the United States did so at the risk of being arrested and sent to federal prison. Companies or individuals engaged in the transfer of biological agents had to be registered with the CDC and exchanges had to be documented. Other nations around the globe had implemented similar controls. Many U.S. researchers considered these regulations a hindrance, an impairment of scientific freedom.

While no one publicly discussed it, some under-the-table exchanges still occurred.

Usually at these conferences, only the stragglers who gathered at the fringes of the meeting rooms and the growing number of law enforcement agents raised any suspicion. Most researchers shared their work freely and stood by to socialize at poster sessions, a long tradition at scientific meetings. In an adult version of the junior-high-school science fair, researchers set up thick white poster boards featuring diagrams, photographs, or blocks of dense scientific text to explain their work. Typically, as colleagues strolled through the conference hall, researchers waited proudly by their displays to answer questions.

The Fourth International, like other such conferences, featured predictable stars, the recognized brain trust of the anthrax world. British-born veterinarian Dr. Martin Hugh-Jones from Louisiana State University used the meeting to update his work with the World Health Organization, where he recorded outbreaks in two hundred different countries. Hugh-Jones, the recognized dean of anthrax experts, had much to reveal about the calendar year 2000, which had brought a spike in reported anthrax events. The Canadians had used helicopters to track an outbreak among wood bison in the Alberta National Park, but the full scale of the epidemic remained unknown. A disturbance of old cattle graves in Nevada had caused three cases of cutaneous anthrax in humans, followed by a livestock outbreak. In faraway Zimbabwe, where anthrax was common, the human cases outnumbered the livestock cases, 960 to 188. The Norwegians had reported a single event of a drug addict dying after injecting himself with anthrax-contaminated heroin. In Japan, anthrax disease suddenly reappeared after nine years of absence.

In his conference report, Hugh-Jones concluded, "we are still hobbled by the Pasteurean belief that anthrax is forever and only control is possible. No. If cases are found promptly, carcasses burnt, not buried, and the stock vaccinated for at least three years, the disease will not just be controlled but eradicated."[6]

From Northern Arizona University in Flagstaff, a team led by the tall, bespectacled Dr. Paul S. Keim, a Hugh-Jones protégé, offered new information about its groundbreaking work mapping the genome of various anthrax strains. Scientific and law enforcement communities closely watched Keim's work at the Flagstaff lab, which collected twelve hundred live anthrax samples from around the globe. Keim had identified the virulent Sterne veterinary vaccine strain used by the Japanese cult Aum Shinrikyo, and helped authorities to trace the strain's source. In the United States, Keim's lab had quashed concern that a cattle anthrax outbreak in California's Santa Clara County could be an act of terrorism. As the worldwide threat of bioterrorism grew, Keim's DNA analysis offered an increasingly valuable tool to law enforcement in locating the source and history of unleashed strains.

Michele Mock led a team from France's prestigious Institut Pasteur that focused on the *B. anthracis* cell structure and function. Anne-Brit Kolsto brought a group from the University of Oslo in Norway that had been conducting genetic analysis of anthrax's close bacterial relatives. The Italians revealed six new anthrax genotypes, and the Chinese reported that anthrax had surfaced in twenty-six provinces, with some strains heretofore unseen in worldwide collections.

A U.S. team from the CDC offered encouraging news about its bioterrorism preparedness and training efforts among state and local health agencies, noting that "an admirable state of readiness for possible bioterrorism threats was achieved . . . in a very short period of time."[7] The CDC was constructing a new state-of-the-art laboratory, where scientists could isolate, identify, and do molecular subtyping of anthrax samples.

The chief of the CDC's Epidemiologic Investigations Laboratory was Tanja Popovic, a dynamic Croatian-born microbiologist with a model's high cheekbones and such impeccable fashion sense that it was hard to envision her waddling around her laboratory in shapeless lab scrubs. She wore painted nails beneath her rubber gloves, remarking when her impossibly long hours kept her from her weekly manicure. Her expertise and dedication to the lab, combined with a sparkling, warm personality, made her a natural at drawing lab technicians in for bioterrorism training. The CDC had trained sixty-four lab techs from fifty states as well as scientists from the FBI to identify *B. anthracis* in field samples, hoping that with specialized training, local lab technicians could quickly recognize the germ, weed out hoaxes, and, when warranted, send suspicious samples up the ladder to the experts at the CDC.

Peter C. B. Turnbull, a world-renowned authority on anthrax disease, came from Britain, where for years he had done seminal research for the public health ministry in collaboration with the defense laboratories at Porton Down. His report to the Fourth International Conference focused on his studies of vaccine effectiveness in cheetahs and rhinoceroses in Etosha National Park in Namibia, Africa. Namibia was a unique natural laboratory for Turnbull's research, since anthrax deaths were a frequent curse among the roaming large animals. Infections among villagers living in primitive conditions also had a long history.

The bookish Turnbull, with gentle eyes framed by dark-rimmed glasses, was endlessly intrigued by what he described as a "beautiful bug" that had been around since antiquity yet remained a "total mystery. We cannot even tell you how an animal gets anthrax today, after all this time. . . . It's just totally fascinating to work with. Anyone who starts to work with it falls in love with it."[8]

Turnbull had taken the lead in organizing the previous two formal anthrax conferences, both in England. He scraped together funding from the British defense ministry and other reluctant sponsors but found little support from public health agencies that considered anthrax too obscure to warrant much investment.

By 2001, Turnbull had cheerfully passed on the conference responsibilities to the Americans from Fort Detrick's USAMRIID, whose sponsorship had been led by Dr. Arthur Friedlander, a prominent anthrax vaccine researcher. From the outset of the conference, Turnbull felt a more aggressive atmosphere, a tension that seemed linked to the increased public awareness of anthrax.

Distinguished scientists from all over the world convened at St. John's College in Annapolis to find that their conference accommodations consisted of overheated student dormitories with old creaky beds. Local bars teemed with rowdy U.S. Naval Academy midshipmen and college students. Even the conference festivities were lacking. During an outdoor cocktail party near the quaint campus boathouse, John Ezzell cringed when his conversation with FBI friends was interrupted by a loud disc jockey who kicked off the evening's entertainment with a raucous tune about drunken partying. Ezzell tapped him on the shoulder and suggested that distinguished international visitors might better appreciate something subtler.

Ezzell had been one of a number of USAMRIID scientists to work with Friedlander on animal tests to gauge the safety and effectiveness of the U.S.–produced anthrax vaccine, a prophylaxis that had been made mandatory for U.S. troops in December 1997 in response to the bioterror threat of the Gulf War. Tall and handsome, with a thick beard and the casual good looks of an aging hippie, Ezzell had grown up a country boy in Concord, North Carolina, and still loved to discuss North Carolina barbecue, even as he traveled around the world to lecture on anthrax. As an adult, Ezzell's easy manner and strong ethical grounding reflected his upbringing.

From a young age, Ezzell wanted to be a veterinarian and forged an enduring relationship with science. As his cousin Larry recalled, John snared a dead bird from a nearby chicken processing plant and used it to place in the school science fair.[9]

"He dissected the chicken, mounted the skeleton on the board, the chicken skeleton that came to be known as George. Then he bought another chicken and took out the heart and the internal organs, put them in baby food bottles and labeled them, put it all in a plastic frame thing, and won second place," Larry Ezzell said.

At North Carolina State University, John took animal studies classes and excelled—not coincidentally—in chicken illnesses. His professor awarded him a pet chicken, which he proudly took back to the Tucker Dormitory. "Tucker's Clucker" became the dorm mascot.

After graduation, Ezzell returned to North Carolina State for his master's degree and became fascinated with laboratory research. He pursued a doctorate in microbiology, then did postdoctoral work concentrating on the germ that causes whooping cough. In 1978, he joined USAMRIID to study Legionnaires' disease. After a fatal outbreak among conventioners at a Philadelphia hotel, the army wanted to analyze its potential threat to U.S. troops.

Despite his scientific credentials, Ezzell seemed an odd fit for USAMRIID, driving in to work many days on his beloved Harley-Davidson wearing a leather jacket and work boots. Armed soldiers in camouflage guarded the entrances of the Fort Detrick building, home of the Department of Defense's lead laboratory for biological warfare research. Ezzell carried a high security clearance—he had been part of a United Nations Special Commission (UNSCOM) team sent to Iraq in 1996 to assess its biowarfare capabilities. After a long motorcycle ride, however, he could look more like a haggard Hell's Angel than one of the army's top experts in bioterrorism.

At USAMRIID's laboratories, a staff of 450, both military and civilian, experiment with the most dangerous pathogens known to man—anthrax, plague, botulism, ricin, equine encephalitis, and the viruses Lassa, Marburg, Ebola, and hemorrhagic fever. With more than ten thousand square feet of isolated space classified at Biosafety Level 4 (BSL4), the highest level of lab containment, it houses the only DOD laboratory equipped to study viruses for which there are no vaccines or treatments. Working in BSL4 is not for the faint-hearted. Lab workers don claustrophobic full-body "space" suits tethered to yellowed, curled hoses for air. Anyone or anything leaving the lab has to be "deconned," or decontaminated, the protective suit washed down in a special full body shower. On a typical workday, BSL4 technicians can be viewed through thick glass panels toiling away under their heavy masks, seemingly unfazed by their occupational hazards. The empty hospital bed in the "Slammer" is used for occasional medical emergencies, but with its crisp white sheets and fluffy pillow, it typically seems little more than a conversation piece for visitors on the official USAMRIID tour.

Since anthrax can be treated with antibiotics and lab workers can be vaccinated against the disease, most of the anthrax work takes place in labs one step down from BSL4. USAMRIID boasts fifty-thousand square feet of restricted BSL3 space, which do not require that technicians don the ominous-looking lab space suits with air supplies. The main risk in a BSL3 lab is exposure to deadly airborne bacteria, so access is tightly monitored. Lab doors carry "danger" signs and airflow is controlled through special "negative air" ventilation systems that protect scientists from contaminants. Most work with pathogens inside BSL3 occurs within biological safety cabinets, which protect both lab personnel and the materials being studied through negative airflow and filtration.

Ezzell had begun his anthrax work after the 1979 accident at Sverdlovsk. The Russian accident's large death toll rocked the Pentagon, where leaders began to fear that U.S. soldiers could be vulnerable to enemy attack. A decision was made to ramp up vaccine studies at USAMRIID, testing a product, then in limited use, that was made by the state-owned Michigan Biological Products Laboratory. Army scientists had shown that the vaccine worked well against the classic Vollum strain of anthrax that had been used in the U.S. bioweapons program, but it remained unclear whether it could withstand a challenge by other, more exotic strains that might be in enemies' arsenals.

"We wanted a strain, a wild-type strain, one that had come from a cow or a deer that had just died of anthrax," Ezzell remembered.[10] A wild strain would test the vaccine's potency. Ezzell's colleague, Greg Knudson, wrote to a U.S. Agricultural Laboratory in Ames, Iowa, seeking such a variety. When the sample came in, the USAMRIID researchers casually named it the Ames strain, not realizing until much later that it had originated in a Texas cattle-grazing field.

In 1988, Friedlander and a USAMRIID team published the results of the first in a series of vaccine "challenge" studies in mice, then moved up the evolutionary chain to guinea pigs. Ezzell and his longtime technician, Terry Abshire, zeroed in on the protective antigens, or antibodies, in animals that were vaccinated, important work that would help to make the vaccines more effective.

Next came monkey studies, which present a number of challenges for researchers. Monkeys kick, bite, and lash out at their captors and must be anesthetized before handling. They quickly learn to recognize and resist needles, forcing scientists to deliver anesthetic with an unwieldy oral tube. Friedlander recalled one important study requiring sixty monkeys the USAMRIID team had to anesthetize twice a day for thirty days. The tests turned into a logistical nightmare, requiring a small army of workers seven days a week: animal caretakers, technicians, anesthesiologists, statisticians, and pathologists to autopsy the monkeys.[11]

USAMRIID used more than a hundred monkeys in the course of its various tests of anthrax vaccines and antibiotics, each animal costing between $2,500 and $5,000, recalled former USAMRIID commander David Franz. Monkey studies were difficult for Franz, trained as a veterinarian, because the primates looked human and glared with sad eyes from their cages.[12]

For USAMRIID's vaccine challenge studies, researchers used a liquid spore preparation of the Ames strain of *B. anthracis.* The spores were suspended in a clear watery solution. The monkeys were anesthetized in their cages inside a BSL3 lab. Once quieted, they were moved to BSL4 for administration of an anthrax aerosol, Friedlander said. Their heads were pushed through tight rubber nozzles into a sealed compartment attached to a nebulizer that turned liquid anthrax into a fine aerosol. Helpless, the monkeys breathed in the deadly mist. Using advanced equipment, the researchers were able to calculate exactly how many spores had made it into their lungs.

USAMRIID's monkey studies gave researchers their first plausible estimates of how many spores would constitute infectious and lethal doses of anthrax in humans.

"In the 1990s, we hired a contractor to go through all that old data to try to help us with things like what really is the dose that it takes to make you sick, to infect you," said Franz. Researchers finally arrived at estimates of at least 2,500 spores to become infected and from 8,000 to 50,000 spores for a fatal dose. These quantities could easily be inhaled in a single breath. "There's a lot of holes in that data; for example, we didn't go down to see the low doses," Franz said. "What if you get a hundred spores? Would any monkeys die, or get infected?"

Years down the road, the questions left untouched by USAMRIID's animal studies would become crucial gaps in the medical world's understanding of the nature and course of human anthrax disease. Law enforcement would look back on the tests for another reason: the 2001 anthrax crimes would raise the sobering possibility that someone involved in USAMRIID's research had gained either the raw materials or the technical expertise to turn anthrax into an aerosol weapon.

In the late 1990s, as public concern about anthrax's potential as a terrorist tool heightened, John Ezzell gradually took on a new role at

USAMRIID. He worked closely with the FBI to train its scientists to detect anthrax while gearing up USAMRIID's laboratories to handle and analyze anthrax that might surface as evidence in a criminal investigation.

Along with his Harley, he kept a four-wheel-drive at the ready at his log cabin on the banks of the Potomac. Single, with two grown children, Ezzell found peace canoeing, hiking, and ambitious gardening. He once bought three thousand flowering bulbs at a clearance sale and planted about 1,000 of them before giving in to exhaustion. He liked to putter through town with George, his chocolate Labrador, panting from the back of his 1952 Chevy pickup. But he also liked to know that he could bolt out of seclusion at a moment's notice if the FBI called in the middle of the night with biological samples to test. He and his lab crew could speed through USAMRIID's gates and be ready for action by the time an FBI chopper landed on the helipad outside his office.

The worst of the anthrax concerns began in February 1998, when Larry Wayne Harris Jr., a certified microbiologist from Lancaster, Ohio, was arrested along with another man in Henderson, Nevada. Guided by a tip that the men were preparing to test pathogens, agents from the FBI's Nevada field office had monitored their conversations and pulled over their luxury car as they drove toward their alleged testing site. The FBI seized the vehicle and whispered ominously to reporters that the men might be carrying enough anthrax to wipe out the entire population of Las Vegas. With great fanfare, the two were charged with possessing an agent of mass destruction.

The suspicion was fueled by Harris's past. Although he was a card-carrying member of the respected American Society for Microbiology (ASM), the outspoken Harris dressed head to toe in black and wore a long, full white beard that dominated his face. He claimed to be a CIA agent, but had been involved with the Ohio Aryan Nation and attended a church known for its anti-Semitic and racist views. Long the subject of law enforcement interest, he had once been questioned about threats against President Clinton. In 1995, he had been convicted of illegally carrying *Yersinia pestis,* an agent of bubonic plague, in his car's glove compartment, after ordering the germs for $240 from the American Type Culture Collection, the same germ house that had filled Iraq's pathogen order. Harris said he experimented with plague and other germs in a home laboratory because he was trying to prove the ease with which anyone could obtain deadly germs. The state charged him with operating a lab without a license, and accused him of mail and wire fraud.

Harris was placed on probation and ordered to perform two-hundred hours of community service, but his scrape with Ohio law enforcement did not silence him. He boasted openly about how easy it was to buy and culture bioterror agents, discussing how anthrax could be procured from tainted cattle graves. He wrote a book about biological warfare, which he sold over the Internet. He considered his work a public service to help Americans survive a biological or chemical attack.

In the 1998 case, the tipster told the FBI that Harris had been retained by a foreign agent to obtain and test weapons of mass destruction. When FBI field agents plugged his name into law enforcement databases, the case took on urgency, recalled Bobby Siller, former FBI special-agent-in-charge of the Las Vegas office.[13]

The impoundment of Harris's car and other evidence presented the FBI with a challenge. The material was potentially dangerous, yet for the purposes of building a criminal case, it had to be protected and quickly tested for authenticity. The FBI did not have on-site equipment, so Siller's agents wrapped the car securely in plastic to prevent spores from leaking and called the U.S. Air Force to airlift it to a nearby base. The other samples were secured and put on a plane bound for Andrews Air Force Base near Washington. From there, they were rushed to USAMRIID for testing by Ezzell and his colleagues in the Diagnostic Systems Division.

At USAMRIID, Ezzell and his team took the samples to his laboratory, smeared cultures of the material on glass slides, and went to work.

Finally, toiling through the night on test after test, Ezzell had the results. He and Franz, then the USAMRIID commander, called the FBI to convey the news, which was passed upstairs to then FBI director Louis Freeh and the office of Attorney General Janet Reno. Some of the material possessed by Larry Wayne Harris was anthrax, but in a harmless veterinary vaccine form. Ezzell heard the phone momentarily go silent at the other end of the line.

There was no criminal evidence on which to hold Harris, who was then in federal custody. Harris said he was released from jail with no explanation around 2 A.M., free to return to his home laboratory: he would not know for some time that his release was thanks to the impartial analysis rendered by USAMRIID's Ezzell and his colleagues.

A few years later, Harris was still speaking at conferences about his bioterror book and how easy it was to acquire biological weapons. At a scientific gathering in New Mexico in 2001, he was autographing books and posing for pictures with admirers when a tall, bearded man approached him.

"Hi. I'm John Ezzell from USAMRIID. . . . I'm the one who tested your anthrax."[14]

Harris, who had since become the only microbiologist ever kicked out of the ASM on ethical grounds, shook Ezzell's hand.

The scientists reminisced about the case and stood together for a photo. Later that night, they met for dinner.

Back at home, Harris continued to preach about the easy availability of germs, this time guiding reporters to a new source of toxins that could be bought, no questions asked, from a culture collection just over the border in Mexico.

In April 1998, joint committees of the U.S. Senate held hearings to revisit the Harris case and assess America's vulnerability to a bioterror attack. Senators from the Intelligence Committee and the Judiciary Subcommittee on Technology and Terrorism questioned Reno and Freeh about potential weaknesses revealed by their response to the Harris incident. No one complained about the FBI's performance or defended Harris's right to possess an innocuous anthrax sample. To the contrary, Nevada Senator Richard Bryan, a Democrat, lauded Agent Siller for averting a potential disaster that could have brought mass devastation to his home state.

Freeh, who had made terrorism preparedness a priority and had activated a special weapons-of-mass-destruction unit at FBI headquarters, told senators that in the three years since a cult threatened to disperse poisonous sarin gas in the New York City subway system, the government had made great strides in preparing for attacks.

"Everything worked the way it should have," Freeh told the panel about the Harris case. His one criticism was that USAMRIID's meticulous anthrax testing had taken thirty hours to procure results. Always anxious for more bioterror funding, Freeh said he wished the FBI had its own mobile laboratory on the ground in Nevada to allow for speedier test results.[15]

Everyone at the Senate hearing realized the Harris case could be just the beginning of a round of domestic terror scares. In the year before the Harris case, Freeh testified, the FBI had investigated 114 bioterror hoaxes around the country, and the numbers were rising. "I'm not trying to paint an overly optimistic picture, but we are much better prepared as a result of the Las Vegas dress rehearsal," he said.

The Harris case only added to USAMRIID's rapidly building workload. Anthrax scares had started in 1997, when chaos broke out at the Washington, D.C., offices of B'nai B'rith after a package was found containing a petri dish labeled "anthrachs." The dish held a red, gelatinous material, and it took about nine hours to determine that it was harmless. The panic and the rush of publicity seemed to inspire a nationwide rash of anthrax hoax letters and packages. The more written about bioterrorism, Ezzell said, "the more the public feared it."

A Christmas Eve threat in 1998 at a California department store caused two hundred shoppers and employees to be needlessly stripped and showered by emergency crews. Hoax letters showed up at the *Washington Post* and NBC's Atlanta news bureau. FBI officials complained that they were servicing hoax alerts every day from some part of the country, including high numbers of threats directed at Planned Parenthood and abortion clinics. "Some abortion clinic would receive a threat letter, and then all of a sudden the antiabortion, pro-life groups were getting threat letters, back and forth," Ezzell said.

Each threat had to be taken seriously, and when samples were retrieved, many of them came by FBI escort to USAMRIID. Ezzell and his fellow lab techs saw everything from tobacco to sand to baby powder pass through the lab, with even the most bogus spreading some level of panic and fear.

Ezzell often scrambled to find lab techs near the site that could speedily evaluate the material. He would counsel them over the telephone through the simplest anthrax tests, telling them to look for the germ's characteristic cloudy suspension in liquids or its rod-shaped form visible under the crudest microscope. "I'd say, 'Tell me, what do you see, what are the physical properties?' and they would describe it, and I would say, 'I don't think it is, but send it on.'

"Sometimes I would be awakened in the middle of the night and someone from the FBI would say, 'Be ready for a conference call in five minutes.' There'd be some incident across the country, and they'd put about ten or twenty of us on the phone at one time. You were being awakened out of dead sleep and asked for guidance," Ezzell said, "and you'd give the guidance. Then the next morning you'd wake up and you were tired and you were thinking, 'What did I say? I sure hope that what I said last night doesn't come back to haunt me.' "

Ezzell didn't mind the telephone consults if they could help the FBI tamp down panic at the scene. "You've got a very volatile situation. People are scared. They're being held. They're being bleached down. They are having their clothes taken off. You had a whole bunch of elementary-school kids one time who were being held until we got some tests back. . . . There was a lot of weight back on us. We were scrambling to handle it."

Over time, the top brass at USAMRIID realized they were no longer running a purely medical laboratory. Environmental samples flooded in from air testing around the sites of the Democratic and Republican national conventions, the presidential inaugural events, and other large public gathering sites where a secret bioterror attack could occur. USAMRIID's mission had increasingly come to include handling forensic evidence with the exacting care and procedures that would allow it to stand up in court. Ezzell spent time with Special Agent David Wilson at FBI headquarters to learn how to develop a USAMRIID Special Pathogens Laboratory that could respond to the need.

"The FBI had been handling occasional cases, like ricin cases, periodically. Back then, ricin was the vogue threat. After Desert Shield / Desert Storm, there was a lot of speculation about anthrax and readiness. But it wasn't until Larry Wayne Harris that for some odd reason, someone got this idea about putting powders into envelopes to scare people. And people around the country saw the reaction—they were fearing anthrax," Ezzell said. "And it became evident very quickly that we needed to have some sort of system across the country so that the FBI could take samples in close proximity to where things were happening."

The truth was that everyone, even the experts grappling with the surge of anthrax hoaxes, viewed the stuffed letters and packages as more a nuisance than a full-blown threat. In pondering a biological attack, they anticipated something much grander in scale and execution, like an aerosol release by airplane or anthrax dropped through a crude bomblet.

As a result, "no effort and no money," Ezzell said, was put into preparing for an anthrax outbreak caused by "something as simple as powder in an envelope."

At the Annapolis conference, Ezzell and Paul Keim presided over a Monday-afternoon panel session on anthrax identification and detection. Friedlander stopped in to hear the presentation, impressed with Ezzell's work.

In many regards, Ezzell was at the top of his game: his reputation was firmly established, his team's research was paving new ground. Somehow, he had managed to blend the excitement of scientific discovery with the drudgery of sorting through the sample backlog then clogging USAMRIID's labs.

But in all of his years of experience, testing phony anthrax samples that turned out to be talcum powder, Ezzell had yet to see *B. anthracis* in its most deadly and scientifically exciting form, the highly dispersible, free-floating weaponized powder produced by Fort Detrick's bioweaponeers before Nixon pulled the plug on germ warfare. Only a few living souls could claim that experience. One of them was William C. Patrick III, the so-called father of weaponized anthrax, who was living in retirement only a few miles from Fort Detrick's tightly secured gates.

CHAPTER 3

Black Clouds over Detrick

1943–2002
FREDERICK, MARYLAND

In a suburban ranch house so distinctive it had once earned a mention in *Better Homes and Gardens* magazine, with floor-to-ceiling glass panels overlooking the countryside near Frederick, Maryland, William Patrick led a visitor down a spiral staircase to his basement office. Certificates and framed newspaper articles hung from the walls, tributes to his years running the product development division of the U.S. Army's offensive bioweapons program. Patrick ranked as one of the world's foremost experts on bioweapons, a fact recognized even by Hollywood. His technical advice helped moviemakers create a goopy biological agent from store-bought orange Metamucil for the 2000 thriller *M-I* [Mission Impossible] *II.* Patrick hated the movie, but his wife loved posing for the picture that hangs above the sofa, showing her arm-in-arm on the set with a beaming Tom Cruise.

From crowded bookshelves, the retired scientist pulled out a thin white binder labeled "Weaponization" in blue Magic Marker. The binder's contents were classified, but Patrick held it aloft to illustrate his point.

"Who in the hell has seen [weaponized] anthrax except Bill

Patrick, Ken Alibek, and a few other people?" he said, laughing dismissively. "Even people like John Ezzell, who are very good scientists, they've never seen any dried anthrax."[16]

Alibek (born Kanatjan Alibekov), a friend of Patrick's, developed biological weapons for the Soviet Union before defecting in 1992 for a consultant's life in the Virginia suburbs. Alibek, who had served as the first deputy director of Biopreparat, the Soviet biological weapons program, supervising thirty-two thousand employees in forty facilities, unloaded a trove of sensitive information to U.S. intelligence agencies when he sought refuge in the United States. No more than fifty other scientists have hands-on knowledge of the weaponization process, Patrick estimated. He had become the self-appointed dean of the group, although some old-timers resented Patrick for what they considered his grandstanding and inflated ego.

Patrick was the creator of several patented processes used at Fort Detrick to turn germs into agents of war. More than thirty years after the program ended, the formulas remained top secret, but Patrick acknowledged that he had discussed some aspects of his work rather freely over the years in private meetings with other scientists and government officials.

Many of Detrick's bioweapons pioneers had retired within the shadow of the fort, their work forgotten by the Pentagon in an ever-shifting array of enemy threats and defense priorities. Some had never again found meaningful scientific work and had never forgiven the government for disbanding a program that they considered vital to long-term national security.

To be viewed with suspicion by the FBI was an indignity. As soon as the first anthrax letter surfaced in October 2001, Patrick expected the FBI to solicit his help in solving one of the most scientifically challenging cases law enforcement had ever faced. Instead, special agents from the Baltimore field office strapped him for three-and-a-half hours to a polygraph machine and grilled him on names, associations, and his movements, phone calls, and consulting jobs, theorizing that he might have wittingly or unwittingly passed on sensitive knowledge to an unknown killer. More pointedly, the FBI wanted to investigate Patrick's association with Dr. Steven J. Hatfill, who had risen to the top of the FBI's list of "persons of interest" in the probe. Hatfill, a brilliant medical doctor who had passed fleetingly through USAMRIID on a research fellowship, had traveled with Patrick, the two sharing the same stage to discuss the bioterror threat. Patrick refused to discuss with anyone but the FBI whether he had ever shared the contents of his weaponization binder with Hatfill.

To demonstrate the dramatic difference between run-of-the-mill anthrax spores and the dried material made through weaponization, Patrick held up two tightly capped glass jars filled with a harmless anthrax simulant. "Watch this," he ordered. The material in the first jar was thick and grainy, settling heavily like table salt after he shook it. The second substance was a white powder finer than confectioners' sugar, so airy that with each slight turn of the bottle it burst into movement and rebounded off the sides of the jar.

"This is a very poor grade of anthrax," Patrick said, again holding up the jar of lumpy material. "You can just look at that and tell that it's granular, no electrostatic charge."

Then he held up the second. "You can look at this and say, 'Gee, this stuff flows like water, small particle size.' And this is what our weapons-grade material looks like."

Patrick's secret formulas were used on real anthrax at Fort Detrick in a twenty-four-hour, seven-day-a-week effort that lasted throughout the 1950s. Under the tightest security, the weaponized powder was shipped to an army production facility in Pine Bluff, Arkansas, loaded into bomblets, and stored. U.S. military leaders hoped never to use the weapons, but they wanted to be prepared. Authorities had incontrovertible proof in the aftermath of World War II that American enemies, especially the Japanese, had developed and tested bioweapons.

President Franklin D. Roosevelt authorized the U.S. biological weapons program in 1942, on the advice of his secretary of war, Henry Stimson. Although Roosevelt and British prime minister Winston Churchill had condemned biological warfare and announced policies limiting it that spring, the president's security advisers warned him privately that U.S. troops needed to be able to protect themselves against enemy bioweapons and to retaliate in kind. Evidence of the Japanese program had been confirmed at the end of World War II, when investigators from what was then Camp Detrick flew to Tokyo to interview captured scientists. Headed by Lieutenant General Shiro Ishii, the Japanese program had included testing of anthrax, botulism, and other agents on prisoners of war, including some from the United States. The Japanese had conducted more than three thousand autopsies on biological test subjects.

During the war, twenty-eight U.S. research universities had quietly worked on bioweapons development, but Stimson considered the program inadequate in light of Japanese progress. Stimson also feared that the Germans had similar biological capabilities, because German saboteurs had infiltrated the United States during World War I and introduced glanders disease into horses and mules. In reality, the Germans were concentrating on chemical weapons, testing their gruesome effectiveness on concentration camp prisoners.

Roosevelt decided to ramp up the biological effort by creating the War Research Service in May 1942, and he chose George W. Merck, the president of the pharmaceutical company that still bears his name, to head the effort. Merck moved to create a centralized biological research program within the army's Chemical Warfare Service. To house the "BWs," or biowarfare programs, the army chose Camp Detrick, an old National Guard airfield conveniently located about thirty miles from Washington in Frederick, Maryland, at the foothills of the Catoctin Mountains. The sleepy village quickly turned into a boomtown.

By 1943, the Detrick research center employed 3,800 military personnel and 100 civilians involved in offensive bioweapons development as well as defensive programs that would help protect U.S. troops, such as research into vaccines for anthrax, Q fever, and other germs believed to be in enemy stockpiles.[17]

Protected by armed guards authorized to shoot on sight, scientists inside the fenced ninety-six-acre compound concocted methods of turning all manner of virulent germs into weaponized material. Early work took place in the Black Maria, a wooden structure covered in tar paper, where botulism toxin was cooked up in ten-gallon tanks.

In 1952, anthrax work moved to Building 470. Inside, army engineers had constructed the massive Eight Ball, a one-million-liter, sphere-shaped aerobiology chamber as tall as a four-story building. The sphere burned irreparably in 1973, but its distinctive design earned it a place in the National Register of Historic Places. Bomblets loaded with anthrax spores could be exploded inside the ball to produce particles small enough to be inhaled. "Over five microns, it would get stuck in the nose. Smaller than one and it comes out like cigarette smoke," explained Joseph Jemski, who ran animal tests in the 1950s.[18]

Researchers testing various pathogens ushered test subjects to stations set up along the exterior of the Eight Ball. Cages of small animals—mice, guinea pigs, and monkeys—were hung from a boom inside the chamber, the animals left to inhale the aerosols for about a minute. Larger animals were fitted with special masks attached to tubing that ran into the sphere.

"We had masks for donkeys, burros, and goats. At first, we wanted dehorned goats. Then we learned that horns were better to tie the masks on," Jemski said.

Gigantic fermenters inside the tall, windowless building cranked out about seven thousand grams of anthrax each week, keeping pace with estimates of Japanese production levels. The anthrax used in Detrick's weapons program was the classic Vollum strain, named for the British scientist who isolated it.

Clearly, laboratory workers risked exposure and death by handling highly infectious anthrax, so the engineering team's first priority was to design new equipment to protect them. Working from scientists' rough drawings, they developed a sealed glass cabinet with built-in gloves that allowed researchers to safely manipulate deadly materials inside the enclosure.

When the War Department publicly acknowledged the biowarfare program in January 1946, it emphasized Detrick's safety record. "As the result of the extraordinary precautions taken, there occurred only 60 cases of proven infection caused by accidental exposure to virulent biological warfare agents which required treatment. Fifty-two of those recovered completely; of the eight cases remaining, all were recovering satisfactorily. There were, in addition to the 60 proven cases, 159 accidental exposures to agents of unknown concentrations," it said in a report.[19]

Danger was an everyday reality. "We all had accidents with anthrax," recounted Patrick, who was twenty-five when he arrived at

Detrick in 1951. In one memorable incident, a can of anthrax spores exploded in the lab, and members of his team were rushed to the infirmary for anthrax vaccine boosters and prescriptions for antibiotics. Then they waited. "And if you had the least amount of sniffle, a little sinus headache, a little cough, 'Oh, my God, I'm gonna get killed by anthrax! I'm gonna die.' That's the black cloud you worked under. Yeah, we had accidents and exposures, but we lived with it," he said.[20]

When Charles Boyles was nine years old, his father, William, a colleague of Patrick's who was assigned to Building 470, took sick about a week before Thanksgiving and closed his bedroom door to suffer in private. Boyles, a fit forty-six, told his family he was coming down with a cold, but in retrospect, he seemed to Charles to know that something was wrong and didn't want his children to worry. The bedroom door stayed shut for several days, and Charles didn't know what to expect. His father, an active outdoorsman and gardener, an avid swimmer, stood at the center of Charles's existence. Charles spent hours staring at the bedroom door, waiting for his dad to come out.

Boyles's flulike symptoms worsened, and around 7 P.M. on November 22, his family doctor admitted him to Frederick Memorial Hospital. At first, Boyles told doctors he didn't think the illness was related to his job, but nonetheless, the doctors felt it best that he transfer immediately to the Detrick infirmary, where doctors were more familiar with exposures to pathogens.

Around 10:30 that night, Boyles checked into the Detrick hospital. The army doctors called in for consultation an expert from the U.S. Marine Hospital in Baltimore, and on the twenty-second, they made a "presumptive diagnosis" of pulmonary anthrax and gave up hope. Meetings were held with the Frederick County public health officer, Dr. Forbes Burgess, to make preparations for Boyles's autopsy and burial. A cloak of secrecy surrounded Boyles's death on November 25, 1951.

Charles Boyles recalled the hushed conversation with Detrick doctors. "I remember they took my mother and sister and me into a room. We were told we were not to talk about it. For years, we did not talk about it."[21]

A press release and death certificate issued on November 26 stated that Boyles died from acute bronchial pneumonia. A few months later, for reasons that remain unclear, Detrick doctors tried to convince the county health commissioner to alter the death certificate to add anthrax as a diagnosis, but the request was rejected. In January 1952, according to an army document, "it was decided that the bacilli found in the brain by examination of fixed slides were sufficiently confirmatory to establish a legally sound basis for a diagnosis of anthrax."[22]

Charles Boyles said his family was never told how William had contracted anthrax, and at the family's request, apparently for insurance purposes, the army produced a letter asserting that Boyles's death did not result from an occupational mishap.

After Boyles's death, Fort Detrick named a street in his honor, the beginning of what became a custom in memorializing biowarfare workers who died in the line of duty. Boyles left another legacy: the army isolated samples of the microbes that killed him and cultured a new substrain of anthrax that became known as Vollum 1B, the *B* representing the last name Boyles.

Patrick lost another colleague in Building 470: Joel Willard, an electrician fell ill on a Sunday afternoon in June 1958, having recently gone into Building 470 to change some lightbulbs. At 9:30 that evening, his wife telephoned their private physician, who suspected a virus and advised him to take aspirin. The next morning, in a panic, she called the Detrick infirmary. Willard was admitted, and blood tests proved positive for *B. anthracis*.

Patrick went to visit him soon after the diagnosis. "Joe was eating, sitting up, and he said, 'Patrick, I'm gonna beat this damn thing.' And the next day he was dead. We don't know how or why he got a dose of spores that would kill him."[23] He died July 5, and another street sign went up.

Because of the anthrax diagnosis, Detrick placed his family on antibiotics, then arranged for the funeral. The body was buried in a lead casket for protection. The undertaker told the family he had never seen such a thing.

In the interest of national security, Fort Detrick withheld the real cause of death. County health officer Dr. Burgess came to a meeting with fort officials that morning.

"After discussion of the security aspects, it was decided the death certificate would specify anthrax as the cause of death, and the newspaper release would specify 'occupational death from respiratory disease,'" Burgess's notes said.[24]

In a photo collection displayed in a less sensitive white binder, Patrick pointed to a grainy black-and-white shot made at Detrick in the 1950s. It showed two young men standing in a portal near the Eight Ball, where they are participating in a human experiment. One man's face is fully protected by an army-issue gas mask; the other wears no protection and stands with his face pressed against a nozzle coming out of the sphere. The masked man looks toward the camera as the other inhales what Patrick calls a "snoutful" of the bacterial agent contained inside the Eight Ball.

Both men were Seventh-Day Adventists, conscientious objectors who, rather than go to war, agreed to take part in an army bioweapons testing program known as Operation Whitecoat, a series of 153 tests conducted between 1954 and 1973. The program, which involved 2,300 men, many of them medics, giving rise to the name Whitecoat, offered American researchers the first detailed evidence of how bacteriological agents move through the human body. Volunteerism was encouraged by church leaders and portrayed as a heroic contribution to the American cause.

Having human "guinea pigs to potential BW agents," Patrick said, was "just unbelievably good for the whole program." The first agent tested in 1954 with a group of about thirty volunteers was Q fever (*Coxiella burnetii*), an illness easily treated with antibiotics. Human reactions were correlated with those of guinea pigs, hamsters, and monkeys. Later, the program tested against tularemia, encephalomyelitis, typhus, and Rocky Mountain spotted fever, all treatable diseases. Jemski, the retired Detrick anthrax researcher, said there was once discussion of exposing the volunteers to aerosolized *B. anthracis,* but because of the exceptional risk, the idea was dismissed and only simulants were used.

"And this program provided tremendous insight," Patrick said. "One, it provided precise data of what constitutes an aerosol dose, and number two, it demonstrated a relationship between dose and particle size."[25]

The program continued through the Vietnam War. Arthur R. Torres, a pastor in Groove, California, signed up in 1963 to avoid going to Vietnam.[26] He spent most of his eighteen-month tour in Detrick's bacteriology lab, near the "hot" areas where anthrax and tularemia were cultured. "We were unprotected from that. . . . We could have very easily been exposed," he said.

In a project to test a vaccine for tularemia, Torres, luckily, was assigned to the control group. After breathing the germs, he was immediately given antibiotics. Some volunteers got very ill, with fevers as high as 105, he said, and treatment was sometimes withheld as long as possible for testing purposes.

Bryon Steele, from Hagerstown, Maryland, now a retired graphic artist, volunteered for Whitecoat projects in 1955. Then twenty-two years old, he was vaccinated against Q fever, then exposed to the germ to test the vaccine's effectiveness. Steele walked into a sealed room, a bit larger than a phone booth, and put on a mask. "A big sphere would blow germs through," he said, and all he had to do was breathe.

In order to ease the psychological challenge of stepping up to the Eight Ball, doctors from Walter Reed Army Hospital recommended a buddy system, Patrick said. The buddies provided kind words and support to the frightened.

After a subject was led to the enormous tank, Patrick said, "You put your snout into an air sampling port. You can't see anything. You can't smell anything. You can't taste anything. And they say you're being exposed to a BW agent. That's scary."[27]

The Whitecoat program ended with no fatalities, Patrick said, and in the late 1990s, former volunteers gathered at Detrick to accept a commendation from the U.S. Congress.

Looking back, Torres said, he was proud of his participation in a program that may have helped protect U.S. troops. But as a pacifist, he regretted that he had not been made "fully aware of all the implications" of the bioweapons program. "I was led to believe it was for defensive purposes only. I've since come to believe that what could be used defensively could also be used offensively."[28]

Along with the human tests, the bioweapons researchers ran a series of eighteen covert experiments to learn more about the potential lethality of germs in the environment. The first large-scale vulnerability study was carried out in San Francisco. In the 1951 experiment, along a two-mile stretch of coastline at the edge of downtown San Francisco, navy ships blasted aerosolized *Bacillus globigii* (BG), which has many of the properties of anthrax but is not infectious. At sundown, researchers would take advantage of the ocean winds to measure how varying meteorological conditions affected a bioweapons release on an open target.

With a "nice ten-mile-an-hour wind," Patrick said, "you're getting more than ten thousand spores per liter. Had you been in that area, I assure you, you would have gotten sick had this *Bacillus globigii* been anthrax."

A second trial using the same amount of bacterial agent, but under less than optimal conditions, yielded far different results. "Unstable air mass, wind was blowing between twenty-five and thirty miles per hour—and we pick up BG only in the first half block of the wharf area of San Francisco," he said.

A similar test measured the release of *Serratia marcescens,* a tularemia simulant. "So we sprayed two miles off the coast, two-mile line, and we only pick up *Serratia marcescens* within fifty meters of the wharf area. The meteorological conditions are good. The aerosol test is taking place at sundown," Patrick said. The researchers concluded that the ultraviolet light was strong enough to kill the vegetative organism, though not strong enough to kill encapsulated anthrax spores.

The test could not be done today, Patrick acknowledged. Scientists have since determined that the organism under the right conditions can infect humans. The government, however, denied responsibility for an infectious outbreak that occurred in San Francisco around the time of the covert testing.

Patrick continued, "When you go to an open-air target like Mother Nature, she really puts some roadblocks in your way. You have to worry about whether you have inversion, you have to worry about wind, ultraviolet light, humidity, temperature." None of those conditions matter when a bioweapon is sprayed into a confined space, he said, "so bioterrorists should limit their activities to enclosed environments."

The attractiveness of enclosed environments was by no means a secret, and in 1966, Detrick researchers undertook a study of the New York City subway system. The results remained classified for many years, so Patrick was stunned when he learned in the late 1990s that the army had released details of the tests to the Church of Scientology.

In the test, the researchers filled lightbulbs with a fluffy anthrax simulant and threw them along the subway tracks. When the train passed, the material flew everywhere. "In a very short time, all the railroad cars had been contaminated and all the subway train stations were filled with this stuff," Patrick said.

The researchers calculated that the average subway rider spent about eight minutes in the bowels of the system, leaving them vulnerable to exposure.

"There are a million people going to work in the morning and in the afternoon, and we would have infected, conservatively, five hundred thousand, and that's very, very conservative thinking. And of those five hundred thousand, probably ninety percent would have died, because you don't know what you're dealing with," Patrick said.

Like the Detrick old-timers, the New York subway study was largely forgotten in the years before bioterrorism became a reality. But by the winter of 2001, the bioweaponeers had proved to be frighteningly prescient. Jaded New Yorkers arrived at subway platforms to find workers in protective suits swabbing the concrete, trying to find out if the anthrax responsible for killing a hospital worker might have wafted through the subway tunnels.

Unlike some Detrick scientists, Patrick developed a brisk second career: he set up a home-based consulting company, Bio Threats Assessment, complete with letterhead that carried the image of the Grim Reaper and a business card featuring a skull and bones. Along with public lectures and frequent appearances on bioterrorism panels across the country, he conducted studies for government agencies and defense contractors, playing off of his Fort Detrick credentials. In government-funded videos to teach first responders how to react to the discovery of a biological agent, he explained in his soft South Carolina accent how to contain an aerosol anthrax release by throwing wet, bleach-soaked towels over the powder.

When researchers at the army's Dugway Proving Ground, seventy miles southwest of Salt Lake City, decided they needed a supply of aerosol anthrax to test detection devices and other equipment, they contacted Patrick.

"They called me up one day and said, 'We want to dry some anthrax, but we don't know how. Can you help us?' And I said, 'Sure. Let's use acetone drying.' So I went out there and we dried *Bacillus globigii*."

Patrick saw nothing questionable about the effort, despite the fact that the United States had agreed to stop making biological agents under the 1972 Biological Weapons Convention treaty.

"They're charged with learning how dry powders can be disseminated. If they don't know how to dry things, how can they test their munitions? For example, during the Persian Gulf War, the Iraqis were lobbing Scud missiles into Israel, and everyone was scared to death that these Scuds contained anthrax.

"So the question comes up, what if you had a Scud and it contains *Bacillus globigii* and it goes *plop,* what is the danger zone? So Dugway has a perfect need to deal with dry powders in order to complete their mission. We were all scared to death that Saddam Hussein was going to lob anthrax into Israel," he said.

Dugway kept details of this production secret until 2001, when authorities began searching for the source of the high-quality anthrax packed into two letters to U.S. senators. Dugway officials then acknowledged producing small quantities of dry anthrax and shipping it in a paste form to a few unspecified locations.

In 1999, Patrick took an assignment offered by Hatfill, who had left Detrick for a job with Science Applications International Corporation (SAIC), a San Diego–based defense contractor with offices in McLean, Virginia. Patrick's assignment was to examine the recent string of anthrax hoax letters and prepare a classified report analyzing the impact a real anthrax letter would have.

Patrick was infuriated by the suggestion, which later surfaced in the news media, that his report could have served as a template for the anthrax crimes. It was hard for him not to feel insulted as the FBI returned several times to his picturesque home.

They came back in the summer of 2002, not to admire Patrick's achievements or consult his notebook on weaponization. Instead, the agents brought with them what Patrick described as two "marvelous bloodhounds" that had recently been used to survey Hatfill's apartment. The dogs were led through Patrick's garage, then came out to the driveway to sniff both the scientist and his wife, looking for any trace of a scent that might somehow be connected to the anthrax crimes.

CHAPTER 4

Enemies Among Us

On or about June 23, 2001
FORT LAUDERDALE, FLORIDA

At first, Dr. Christos Tsonas saw nothing unusual about the two well-dressed Middle Eastern men who showed up in the Holy Cross Hospital Emergency Room. Tourists, he thought. The men, however, told him they were airline pilots living on Bougainvillea Drive in Lauderdale-by-the-Sea, an oceanfront town popular among scuba divers for the beautiful reef only a hundred yards offshore. The older man did most of the talking for his roommate, a slender, dark-skinned man in his early twenties. The young man had torn open his leg several months earlier by bumping into the edge of a suitcase, the interpreter explained, and had developed a nasty, inch-wide ulcer on his leg. The injury was not healing naturally. Their landlord, who rented them a small apartment attached to a private home, had seen the wound and recommended a visit to Holy Cross, a nearby Catholic hospital sponsored by the Sisters of Mercy.[29]

A dry blackish scab covered the wound, the edges swollen and rimmed in red. Tsonas examined it curiously, removed the scab, cleaned the wound thoroughly, and wrote a prescription for Keflex, an antibiotic commonly used for bacterial infections. The whole encounter took about ten minutes. No medical alarms went off. He did not culture the tissue or recommend further tests, and the two men went silently on their way.

It was June 2001: John Ezzell and the world's anthrax experts were calmly discussing anthrax science in Maryland, and there was no reason for a harried emergency medical specialist like Tsonas to link the man's symptoms to a disease rarely seen in Florida. No alert had come down from the nation's public health agencies, which were quick to notify emergency rooms about unusual disease outbreaks, like the mosquito-borne West Nile virus that was showing up that summer in scattered locations along the East Coast.

Tsonas did not think about the lesion until four months later, when FBI agents showed up trying to trace medication prescribed to one of the 9/11 hijackers. They had found the prescription during a search of the residence shared by Ziad Samir Jarrah and Ahmed Ibrahim al-Haznawi, who had both been aboard the San Francisco–bound United Airlines Flight 93, which was hijacked after a Newark, New Jersey, takeoff and steered toward an unknown destination in Washington. After a passenger uprising, the Boeing 757, the last in a line of four jetliners hijacked that day, had crashed in a field in Somerset County, Pennsylvania, killing all forty-four on board. Authorities believed that Jarrah piloted the plane while al-Haznawi and three other al-Qaeda foot soldiers tried to subdue the crew and passengers.

The federal agents pulled out photos from recently issued Florida driver's licenses and Tsonas identified the pair: Jarrah, with his slight friendly smile, and al-Haznawi, serious and frail looking. Then the FBI gave Tsonas his own medical record of al-Haznawi's emergency room treatment, and an uneasy realization flooded over him.

His write-up of al-Haznawi's blackish wound, written in the calm days before bioterrorism became an American reality, was a textbook description of cutaneous anthrax. And skin anthrax could be acquired in only one way: through direct contact with anthrax spores. For the first time that week, clues about the hijackers' movements in South Florida and elsewhere in the months before their suicide mission raised a grim possibility: the terrorists who had wreaked havoc on New York and Washington also might have been planning an anthrax attack on U.S. citizens.

If proven, the theory might have been a relief to the overburdened FBI, since it was easy to imagine the same band of murderers plotting with accomplices to unleash another round of terror while Americans were still reeling from the 9/11 attacks. The references to Allah and other tip-offs in the anthrax letters suggested Muslim origins, which seemed to strengthen arguments for ties to al-Qaeda. The central unanswered question in this theory was how the terrorists had managed to secure their anthrax samples, especially since DNA analysis showed that the anthrax in the letters matched the strain used for U.S. military research. Iraq, which was considered the most likely base of an al-Qaeda operation, had tried to get the Ames strain in the 1980s but was thought to have been unsuccessful—at least through legitimate channels. Many bioterror experts argued, however, that given the loose controls, Iraq could have secretly acquired the Ames strain in any of dozens of ways and could have cultivated it by the boxcar-load by the time the U.S. attacks occurred.

In the frantic weeks after 9/11, FBI agents scoured the cars, apartments, and possessions of all of the suspected terrorists. When the trail of anthrax letters began showing up in October, the bureau told the public that thorough tests for signs of anthrax had come up empty. The special agent heading the Miami field office, Hector Pesquera, tried to control speculation that the hijackers had been behind the anthrax mailings, telling journalists it was no time for "premature conclusions and inaccurate reporting."[30]

Yet the October visit to Tsonas was one of dozens of FBI interviews conducted in Florida to explore a series of inconclusive clues that the 9/11 hijackers were preparing a biological weapon for use in the United States. A top FBI official insisted that agents on the anthrax probe worked hand-in-glove with investigators on the al-Qaeda hijacking case to explore any possible links.

Some of the 9/11 hijackers had trained in al-Qaeda camps in Afghanistan, and U.S. intelligence had documented al-Qaeda's keen interest in biological agents. There were inconclusive reports that before 9/11, Mohammed Atta, the alleged ringleader of the U.S. hijacker operation, had met in Prague with an Iraqi intelligence case agent or liaison, Ahmad Khalil Ibrahim Samir al-Ani. Reports indicated that Czech intelligence officers had seen Atta accepting a glass container during one meeting, leading to speculation that he might have acquired an Iraqi anthrax sample.[31]

Indisputably, Atta had shown an unusual interest in the crop dusters used to spread fertilizer and chemicals on Florida's expansive farmlands. Experts in biological agents had long described such low-flying aircraft as an ideal means to achieve maximum effectiveness in disseminating an aerosol over a city.

Atta, while living in South Florida, had made two suspicious visits in 2001 to check out crop-dusting equipment. He had his eye on the Air Tractor AT-502, worth about $50,000. During his first trip to a rural airstrip, Atta had driven up at the end of a workday while crewmember James Lester washed a dirty airplane, covered in dry fertilizer. Atta asked pointed questions: How much could the plane carry? How did the engine start? He wanted to crawl into the cockpit.

Lester got so annoyed by the pushy foreigner that he asked him to leave. "I refused to tell him how to start it," he said. "Eventually I had to push him away. He was walking right at my back. I told him, 'I've got no authority to open this cockpit and let you get on this airplane!'"[32]

Atta returned several weeks later, this time with two companions and a camera. They snapped shots of the plane and walked around its perimeter while Lester worked on changing a tire. Lester glared at them, making eye contact with Atta, and they did not approach, walking around the field and eventually driving away.

When the FBI showed up with mug shots a few days after 9/11, Lester had no problem picking Atta out of the collection. "I was standing toe-to-toe with the guy," he recalled. "It wasn't like you could forget him."

Atta also had made inquiries at the U.S. Department of Agriculture's Farm Credit Services about getting a government loan to buy a crop duster. It would have been yet another manipulation of the U.S. system—winning federal backing to buy a plane that could spew chemical or biological weapons. But the USDA employees explained that the agency made no such loans and referred Atta to a commercial bank. The FBI interviewed officials of the Community Bank of Florida about Atta's loan inquiries.

Atta showed up at Huber Drugs in Delray Beach looking for medicine to treat a red rash on his arm. The pharmacist recommended a wash for acid burns. Later, he reported the encounter to the FBI after the first anthrax attack, only a few miles away, wondering if Atta's irritation could have been caused by bleach, which is used for decontamination of anthrax accidents.

Atta was in South Florida at a time when thirteen of the nineteen terrorists came together in the area, training for strength and stamina in local gyms while working out the elaborate details of the 9/11 plot. Most of the plan's foot soldiers, including al-Haznawi, flew in from foreign countries between April and late June. Jarrah and the other designated pilots for the operation had come much earlier for flight training in various locations.

Jarrah, Atta, and Marwan al-Shehhi had begun their association as members of an Atta-led al-Qaeda cell based in an apartment in Hamburg, Germany. After the 9/11 plot was hatched, the three came to the United States around the same time to start flight training. Jarrah soon tried to enroll another Hamburg cell member, Ramzi bin al-Shibh, in his Florida flight school, but al-Shibh, who is from Yemen, a hotbed of terrorism, failed in four attempts to get a U.S. visa. The FBI later conjectured that bin al-Shibh was intended to be part of the plot, and when he was denied entry, a French-Moroccan student named Zacarias Moussaoui was later dispatched in his stead to be the twentieth man on the hijackers' roster.

If Moussaoui intended to participate, which he strenuously denied, he never got the chance. When he enrolled in a Minnesota flight school for instruction in flying a 747 airliner, he aroused suspicion by telling instructors he did not need to know how to take off or land, only how to keep the plane on a steady course. He, too, seemed interested in crop dusters and asked questions about them. Alerted by the flight school, federal authorities detained him on immigration charges, and he was in custody at the time of the 9/11 attacks. Moussaoui was indicted in December 2001 and brought up on charges that he plotted with the 9/11 hijackers.

Jarrah's involvement in the plot puzzled members of his family, who insisted that he had never been an Islamic zealot. He loved Western culture and enjoyed dancing and socializing, his mother told interviewers. He had a serious girlfriend at the American University in Beirut, a school patronized mainly by the children of the westernized elite of the Arab world. The couple had planned to marry in the fall of 2001.

Upon arriving in the United States, Jarrah settled comfortably into the American lifestyle. He enrolled in a gym near Fort Lauderdale and hired a personal trainer, diligently working out two or three times a week but still not impressing his instructor. He and Atta got their Florida driver's licenses on the same day, May 2, 2001. Atta had already chalked up a speeding ticket in his red 1986 Pontiac, and had a court date to show up in late May with a valid Florida license, which he ignored.

Born October 11, 1980, al-Haznawi came from Saudi Arabia, where his father was an imam in the village of Hezna. His desire to be part of the Islamic jihad had caused a falling-out with his father, and he had left home without permission in 2000 to fight in Chechnya, urging distant cousins to join him. Two of his relatives, Ahmen and Hamza Alghamdi, were aboard one of the jetliners that plowed into the World Trade Center.

After his arrival in Florida, al-Haznawi and other newcomers to the terrorist operation set up bank accounts and visited the Department of Motor Vehicles. They played like rambunctious college kids on the area's wide beaches. Except for the June emergency room visit, nothing about al-Haznawi raised suspicion.

It is unclear why Jarrah took al-Haznawi under his wing or why the two moved together into the $200-a-week flat in Lauderdale-by-the-Sea. Other hijackers were living nearby in several apartments that al-Shehhi had arranged through a local realtor, Gloria Irish, who happened to be the wife of an editor at the *Sun,* a publication of American Media Inc., the same Boca Raton media company hit first by the anthrax attacks. The hijackers' involvement with Irish was yet another connection that the FBI would eventually judge to be a bizarre coincidence.

As summer rolled by, the 9/11 operation moved into high gear. Jarrah left the United States briefly, and members of the terrorist group began fanning out along the East Coast to situate themselves near the airports where they would carry out their missions.

On August 29, 2001, Ahmed al-Haznawi purchased his one-way ticket for a September 11 journey on United Airlines Flight 93 from Newark to San Francisco. The other hijackers did the same, some using the Internet to place their orders and even entering frequent flier numbers to show their bona fides. The final phase of the plot proceeded without a hitch. In the words of a federal indictment,

> *On or about September 11, 2001, Saeed Alghamdi, Ahmed Alnami, Ahmed al Haznawi, and Ziad Jarrah hijacked United Airlines Flight 93, a Boeing 757, which had departed from Newark, New Jersey bound for San Francisco at approximately 8:00 A.M. After resistance by the passengers, Flight 93 crashed in Somerset County, Pennsylvania at approximately 10:10 A.M., killing all on board.*[33]

Al-Haznawi remained a bit player in the coverage of the 9/11 conspiracy until months after the attacks. The following spring, the Arab television network al-Jazeera, which had become the conduit for videos featuring pronouncements by al-Qaeda leader Osama bin Laden, released a tape purported to be al-Haznawi's last will and testament. A disturbing image of the blazing World Trade Center was superimposed over the background.

Dressed in camouflage, his head draped in a black-and-white kaffiyeh, or Arab scarf, a young man believed to be al-Haznawi delivered a virulent anti-American attack, then offered himself up as a sacrifice for al-Qaeda's cause. The al-Jazeera representatives said the tape had been filmed six months before 9/11 in the Afghan city of Kandahar, where al-Haznawi had apparently gone for training before embarking for Florida. It was the first time Americans had heard one of the nineteen hijackers speaking in his own voice.

"The time of humiliation and subjugation is over. It's time to kill Americans in their heartland," he proclaimed. "Lord, I regard myself as a martyr for you to accept me as such."[34]

In early 2002, Tsonas once again reviewed his skimpy notes from al-Haznawi's visit, trying to recall clouded memories of the brief emergency-room visit. Doctors from the Johns Hopkins University Center for Civilian Biodefense Studies had contacted him when federal health officials asked the institute to determine the lesion's probable cause. The assignment presented many difficulties. Their findings would be too insubstantial to publish in a peer-reviewed medical journal; they would be only an opinion, perhaps worth further consideration as the investigation continued.

The center's doctors, Tara O'Toole and Thomas Inglesby, studied the evidence and gathered feedback from other medical experts at John Hopkins, including a dermatologist. They researched the medical literature and textbooks for any possible causes of a black eschar with raised edges and explored whether the wound could have been caused by an infected spider bite. Al-Haznawi's suitcase-bump explanation was summarily dismissed. In the end, the Hopkins group wrote a short memo concluding that the lesion was "highly suspicious for cutaneous anthrax," but, as O'Toole said, "there was no smoking gun."

Although the panel's review was strictly medical and offered no opinion about the hijackers' involvement with biological agents, a *New York Times* report based on the study touched off a flurry of new speculation about the source of the anthrax mailings.[35] O'Toole told the *Times* there was "legitimate concern" that the FBI might not have the technical expertise to put together fragments of important information.

The FBI, however, dismissed the conclusion of the Hopkins panel. "This was fully investigated and widely vetted among multiple agencies several months ago," said spokesman John Collingwood. "Exhaustive testing did not support that anthrax was present anywhere the hijackers had been. While we always welcome new information, nothing new has in fact developed."[36]

Collingwood's remarks hardly inspired confidence in the *Sun* and *National Enquirer* newsrooms, where employees were mourning the loss of Bob Stevens. The first anthrax attack in America had occurred within a short drive of the hijackers' temporary home. The string of coincidences linking the hijackers to anthrax was too compelling to ignore. Months after reporters evacuated their offices in a panic and lined up for Cipro, veteran reporter Don Gentile of the *Enquirer* chuckled cynically when asked if he thought the hijackers could have orchestrated the anthrax letters before plunging to their deaths. "Yeah," he said softly. "I believe they had something to do with it. I'll always believe it."[37]

CHAPTER 5

Walking Wounded

October 4, 2001
NEW YORK CITY

Dr. Jeffrey Koplan, director of the U.S. Centers for Disease Control, had been rushing to catch a plane at LaGuardia Airport when his cellphone rang with an urgent call from his Atlanta office. "Get back to us on a secure line," a nervous aide instructed.

Koplan resigned himself to more bad news. He had come to Manhattan on crutches, his foot recovering from extensive surgery, to check on the CDC operation mobilized there in the aftermath of 9/11. Seventy CDC staffers were working around the clock on recovery efforts near ground zero. Among their many missions was helping overwhelmed city health authorities conduct an investigation of hospital emergency rooms, looking for strange illnesses that might suggest the release of a biological agent. Over the years, health officials contemplating the terrorist threat had often speculated that a bold attack like the bombing of a building might be followed by a second, more secretive bioterror assault, using anthrax or another pathogen. But the CDC's hospital review, conducted with slips of paper since Manhattan's computer networks were down after the attacks, had shown no spikes in the number of infections or other ailments that could suggest contact with pathogens.

From LaGuardia, Koplan found a secure telephone and called his office back for a briefing. The CDC had received word the previous day of a patient in Palm Beach County, Florida, who was rapidly declining from a mysterious illness. A normally healthy man, the patient had become severely ill after returning from a North Carolina vacation with his wife. He had been admitted to the hospital with a 102.5-degree fever, lethargic and disoriented. The initial diagnosis was spinal meningitis, a brain inflammation. Within hours, he had suffered a grand mal seizure and was in a coma, his breathing controlled by a ventilator. An initial chest X ray had shown an odd widening of the space between the lungs.

Infectious disease specialist Larry M. Bush, the patient's doctor at the John F. Kennedy Medical Center, had become alarmed when he examined a sample of spinal fluid and did a standard gram-stain test to identify the type of bacteria causing infection. Under the microscope, Bush saw a profusion of bacteria in the dying man's bloodstream, the germs resembling the rod-shaped members of the *Bacilla* family. The gram-stain test further narrowed the possibilities. Doused in purple dye, the sample did not turn pink but retained the purple color, telling Bush that it was among a select group of "gram-positive" bacteria. Though he had only studied the germ in textbooks, Bush became convinced he was dealing with *B. anthracis.*

He had called Jean Marie Malecki, the county health department director, and asked her to sit down before sharing his analysis. Malecki had asked that a sample be sent immediately to Florida's state health laboratory in Jacksonville for preliminary analysis, which confirmed Dr. Bush's findings.

Florida-based FBI agents had driven all night to bring the sample to the CDC for further testing. In Koplan's absence, James Hughes, director of the CDC's National Center for Infectious Diseases, and Bradley Perkins, the special pathogens division chief, had alerted lab director Tanja Popovic at home just as she sat down to dinner. Popovic put in a call to Phillip Lee, a Florida State lab technician who had only recently completed Popovic's CDC training program on bioterror agents.

Lee described his testing procedure, and Popovic reported back to her bosses that he had followed all the right steps, although it was impossible for her to make a definitive finding for anthrax over the telephone. She wanted the samples brought to Atlanta, where cultures could be grown and analyzed.

Two rapid confirmatory tests at the CDC labs were positive for *B. anthracis.* The cultures would be ready later that morning.

Koplan, tall and thin at fifty-six, maneuvered down the airplane's narrow aisles and buckled up for the flight to Georgia. The report from his office mystified him. Authorities had braced for a possible act of bioterrorism, but they had focused on New York.

How could anthrax, rarely seen along the East Coast, be killing a man in Florida?

Since 1998, when he accepted an invitation from President Clinton's health secretary Donna Shalala, to serve as CDC director, Koplan had placed heightened emphasis on preparing for bioterrorism. The challenges were daunting. A CDC study of the nation's public health clinics, laboratories, and hospitals conducted in 1999 showed that they were astonishingly unprepared to react to a bioterror attack. Only 45 percent of the local health departments that responded to the CDC's survey had the ability to send simultaneous faxes in the event of an emergency. Less than half had Internet access, and twenty percent had no e-mail capability. One local health department confessed to not reporting diseases to federal authorities to avoid the cost of a long-distance phone call.

Koplan had guided a $130 million federal effort to improve communication and data collection and to train first responders. A national laboratory network was rapidly being developed so that if bioterrorists struck in remote locations, local authorities could analyze the agents used and the extent of the threat.

The CDC had also beefed up its own ability to test and analyze pathogens sent in from the field. The new Epidemiologic Investigations Laboratory headed by Popovic, boasted BSL3-plus containment labs and DNA sequencing machines that could help identify a pathogen's source.

With Shalala's departure after the 2000 elections, Koplan had kept the effort on course under a Republican administration with different spending priorities. The administration had suggested deep cuts in bioterrorism preparedness spending, and Koplan feared that it would also slash funding for other CDC initiatives.

As Koplan's boss, President Bush had appointed Tommy Thompson, the former Republican governor of Wisconsin, to serve as secretary of health and human services. Billed a "compassionate conservative," Thompson had won national attention as the architect of Wisconsin's groundbreaking welfare-to-work legislation. During his January 2001 confirmation hearings, Thompson had vowed to double funding of the Bethesda-based National Institutes of Health (NIH), the respected federal medical research institute. He had said little about his goals for the CDC, even when he made his first visit to the Atlanta campus that spring. Some of the CDC's health initiatives had come under harsh criticism from conservative groups. On the Atlanta campus, Koplan had proposed a costly capital improvements program to upgrade outdated laboratories and shabby buildings. Thompson had slashed Koplan's proposed funding in early budget discussions on Capitol Hill.

Koplan's friend Dr. Philip Brachman recalled that in earlier days, CDC officials loved the autonomy that came with being the only federal agency not located in Washington, D.C.[38] But Thompson's strong, centralized managerial style, a sharp departure from Shalala's, had highlighted the disadvantages of the CDC's remote location. It was hard enough, even during the best of times, to handle delicate, politically charged relationships with multiple agencies from a distant federal outpost. The events of 9/11 and its aftermath, with soaring demands for medical advice and services, would strain the relationship between the experts in Atlanta and the politicians in Washington to the point of total breakdown.

Since medical school, Koplan had felt a deep affection for the CDC and its public health mission. He was a resident at New York's Montefiore Hospital in the early 1970s when Brachman, a noted infectious disease specialist with what was then called the Communicable Disease Center in Atlanta, visited New York and urged Koplan to use his student military deferral to join the

Epidemic Intelligence Service, an arm of the CDC dating back to the Korean War. The young doctor took him up on the offer.

Arriving at the CDC in April 1972, Koplan became part of a worldwide effort to eradicate smallpox. He embarked on a clinical trial of a new smallpox drug in Bangladesh. Koplan took over an empty hospital ward in Dhaka, hired a former Pakistani army sergeant as his assistant, and moved in patients whose rashes would often progress to scarring or death. Faces swollen and distorted beyond recognition, many deteriorated quickly and died within days. The drug Koplan tested proved to be ineffective, but the experience was invaluable, particularly when smallpox emerged in the late twentieth century as another potential terrorist tool. Few doctors in the United States had hands-on experience caring for smallpox patients.

After finishing his training in internal medicine, Koplan was dispatched by the CDC to the West Indies, an inspiring time during which he learned to play a steel drum and developed a passion for the Caribbean. Koplan, who had studied poetry as a Yale undergraduate, indulged his writerly instincts with prolific writings for medical journals, including one in the *Lancet* questioning the value of stool samples. The prestigious journal rejected Koplan's creative title, "The Incremental Value of Excremental Testing."

In December 1984, he led a team of U.S. scientists and academics to India in response to a chemical disaster at the Union Carbide–owned pesticide plant in Bhopal, with 2,000 casualties and 100,000 injuries. Walking past cremation pyres and through slums where toxic fumes had blinded masses of peasants, Koplan witnessed the horror of mass devastation.

After five years at CDC headquarters, Koplan took a private sector job, heading the Prudential Center for Health Care Research. But when Shalala asked him to reenter the lower-paying, infinitely more demanding public health arena as CDC director, he jumped at the chance. The job appealed to Koplan's sense of mission. Working with talented doctors and staff, he believed that directing the agency was not just professionally challenging but important to the betterment of the nation's public health system.

Reporting directly to the HHS secretary, the CDC director oversees a dozen centers and institutes and more than 8,000 employees,

5,600 of them in Atlanta. The agency is so spread out over twenty-four Atlanta locations that shuttle buses cart workers from one flat, 1960s-era brick building to another.

Founded in 1946, an outgrowth of a World War II agency charged with fighting malaria, the CDC had grown to a $4-billion-a-year agency. On a typical day, Koplan held back-to-back staff meetings, conference calls with HHS officials in Washington, and hurried telephone interviews with the press on an endless array of health questions. Even his drive to and from work usually included a scheduled cellphone call to catch up on some bit of important business.

He also gave frequent briefings on public health issues: HIV-AIDS, West Nile virus, obesity, and tobacco-abuse prevention. A dedicated three-times-a-week swimmer who liked to row on the Chattahoochee River, Koplan, a father of two, had pressed Americans to adopt healthier lifestyles with exercise and a balanced diet.

The events of 9/11 had shown the CDC staff functioning at its highest level, Koplan believed. Within hours of the New York attacks, Koplan, in consultation with Thompson, had dispatched the first push-pack of supplies from the National Pharmaceutical Stockpile and a team of six CDC staffers to administer it. Filled with such essentials as antibiotics, vaccines, intravenous tubes, and face masks, officials had estimated that the push-pack could be dispatched to the site of a biological or chemical emergency within twelve hours. Through decisive action, the supplies had made it to New York through restricted airspace in just four hours.

The subsequent hijacker attack on the Pentagon in Washington only increased the pace in Atlanta. Then, around 10 A.M. that morning, through a string of communications with the FBI, Koplan learned that a passenger plane had left a Georgia airport heading north to an unknown destination. U.S. intelligence had received an unspecified threat that the plane could be headed for the CDC.

Koplan wasn't one to give in to the panic of hurricane alerts or tornado warnings, and the notion of a hijacker zeroing in on the CDC campus seemed preposterous to him. He looked around the room, every person responding to an urgent need, and could not imagine interrupting the work for a likely false alarm.

Then reality sank in. The morning's events had proved that he could not disregard an official warning. Although the CDC had a low profile, it did, in fact, house an extensive inventory of pathogens, including one of the world's few remaining samples of the smallpox virus. Koplan gave the order to evacuate, dispatching key staff to an off-site emergency command center in nearby Chamblee.

When his wife, Carol, a child psychiatrist, called him from home, Koplan asked her to meet him in their driveway with a clean shirt. Barely slowing down, he swooped by the house, picked it up, and asked her not to wait up for him. It was the first of many sleepless nights.

The evacuation lasted only a day. Officials moved back into headquarters, and contemplated the possibility that the hijackings could be an initial round in a larger calamity, possibly involving chemical or biological terrorism. As Dr. Stephen M. Ostroff, the associate director for epidemiologic science, explained, the theory was "if you're going to do something like that, the best time to do it is hit them when they're down."[39]

Still, no one was prepared for the call from Florida. The hospitalized patient, sixty-three-year-old Bob Stevens, was a photo editor who lived in Lantana and commuted each day to work in Boca Raton. But by the time Koplan arrived back in the office from New York, Popovic's cultures were in: Stevens definitely had anthrax. Hughes had asked Perkins to head the CDC team leaving for Florida to help with the investigation. A chartered jet waited at Peachtree-Dekalb Airport to whisk them to Palm Beach County.

Their first stop was the Xerox machine. The U.S. had not had a case of inhalation anthrax since 1976, and medical literature on human cases was surprisingly thin. Perkins, who had never examined an anthrax patient, wanted to use the short flight to hurriedly cram before beginning work on Stevens's case.

The most relevant twentieth-century studies had been done by Brachman, who investigated an anthrax outbreak in 1957 at Arms Textile Mill, a goat-hair processing plant in Manchester, New Hampshire. Only one of five victims in the outbreak had survived. The case gave U.S. epidemiologists a chance to study the opportunistic nature of the disease and to explore why some workers contracted it while others stationed near them were spared. The inhalation victims had worked in a baffling variety of areas; some operated wool-combing machines, while others worked in the carding or weaving departments. Brachman discovered the common denominator: all of the victims had handled a particular shipment of black goat hair from Pakistan. Anthrax spores filled the packages.[40]

Perkins's team had headed to Florida to tackle a more challenging mystery. How had a healthy man who spent most of his time sorting photographs and making telephone calls from an office desk in Boca Raton inhaled the virulent bacteria? Had something unusual interrupted his routine, exposing him to the germ?

One CDC team already had embarked for North Carolina to retrace Stevens's steps during a brief family vacation, and Perkins's Florida group would focus on his home and hangouts in Lantana and Boca Raton. If they failed to prove that the bacteria had come from a natural source, they would have to address the probability that the case was an attempted murder—possibly an act of bioterrorism. Perkins wondered if Stevens could be the "index case," the first patient in a wider outbreak. How many others could be lying in Florida-area hospitals, misdiagnosed and untreated?[41]

As Perkins flew into South Florida, it was hard to ignore the footprints of the 9/11 hijackers. Florida FBI agents were already examining possible connections to the terrorists.

For Perkins, looking down from the air to see spots highlighted on a map of South Florida, the geographic connections seemed too striking to be mere coincidence.

The Attacks

CHAPTER 6

The Index Case

October 4, 2001
PALM BEACH COUNTY, FLORIDA

Bradley Perkins's team joined Malecki's county health department squad at a new emergency center originally built for hurricane relief. The center's phone banks and computers had been used only twice, once to coordinate response to the devastating Hurricane Andrew and a second time to re-count dimpled ballots in the whisker-close presidential elections of 2000. Along with doctors, lab technicians, and pathologists, the group assembled to investigate Stevens's case included John Bellamy, a Miami agent assigned to the FBI's weapons-of-mass-destruction unit, and, oddly enough, a U.S. postal inspector, although no one at the time had any inkling that the anthrax infecting Stevens had come through the mail. Because most recent anthrax hoaxes had involved mailed letters, the FBI had begun to bring postal inspectors along in emergency responses to lend their expertise in identifying and tracking suspicious mail.

The response group hurriedly set up overnight camp, the Palm Beach squad on one side of the room, the federal newcomers on the other. By morning, they would embark on a common mission: to examine each piece of Stevens's recent life, track his movements, interview his friends and coworkers. For guidance through the details of his daily life, they had to rely on Stevens's family, because he was unconscious after his grand mal seizure.

Perkins decided it would not be necessary to examine Stevens extensively. He trusted Larry Bush's findings and approved of the drug regimen put into place for Stevens's treatment. Still, he badly wanted to see the patient himself. The prospect of examining a living patient felled by inhalation anthrax intrigued the young physician. Perkins, whose specialty was infectious diseases, knew he might never again get the chance to see this disease in action. Even Dr. Friedlander, USAMRIID's veteran anthrax expert, had never examined a human inhalation anthrax patient.

When Perkins arrived at the 370-bed JFK Medical Center in Atlantis, Florida, he found Stevens's wife, Maureen, and his immediate family holding vigil in a hallway. They were distressed that yet another doctor had arrived to examine their ailing loved one. Maureen had driven Stevens to the emergency room, perturbed that he had not heeded her advice to stop for medical treatment when he began feeling ill in North Carolina.

Terrified by Stevens's sudden, inexplicable decline, the family confronted Perkins, desperate for reassurance.

In a Midwestern monotone that suggested his Kansas roots, Perkins told the family that Stevens seemed to be responding to the battery of powerful antibiotics. The doctors were doing all that they could. But he knew that Stevens's chances of recovery were slim to none.

Perkins ducked into the patient's room and stood for a few minutes by the bedside, calling Stevens's name. There was no response. He examined the medical chart hanging by the bed. The patient had a history of mild hypertension and had a stent in his heart to repair damage from a mild attack, but otherwise he was a healthy nonsmoker. When he arrived at the hospital, he had a fever, normal blood pressure, and high white blood-cell count.[42]

Perkins heard the patient's labored breathing through the ventilator. A screen displayed weakening vital signs. The culture of Stevens's spinal fluid left no doubt that anthrax bacteria were permeating his system. The index patient was dying. Now the challenge would be figuring out whether his exposure was a fatal mishap or an act of cold-blooded murder.

Still in pain from recent gallbladder surgery, Jean Malecki had barely slept since Dr. Bush phoned her from JFK to report the unusual bacteria growing in Stevens's spinal fluid. He had called while she was conducting a biological and chemical training seminar for about eighty medical professionals from fourteen hospitals. Bush, an old friend from her days practicing at JFK, left a message at her office for her to call him back immediately.

"You have to keep this very quiet," she heard him whisper into the telephone. Bush had described his patient and the disturbing results of a gram stain of the spinal fluid. As unlikely as it seemed, the germ looked distinctly like *B. anthracis*.[43]

Malecki urged caution as she sank wearily into her office chair. Her stomach hurt, and she could feel this new wave of stress tangling it in knots. How unlucky could she be to end up with a case of anthrax in her jurisdiction?

Unlike Jeffrey Koplan, with his well-appointed executive space at the CDC, Malecki ran her public health empire out of a spartan office with institutional furniture and thin wall-to-wall carpeting. Malecki served a population of more than a million, her job included duties ranging from running school immunization and maternity services to administering seven public health clinics and a wide range of environmental programs. Despite the job's demands, Malecki valued her role on the front lines of the public health system and had concentrated on the new threats of the bioterror age more than most local health officials across the country. The first step, she told Bush calmly, would be more testing on Stevens to try to eliminate anthrax as the agent. Maybe the gram stain was wrong. If further tests still suggested anthrax, she said, then the doctors could "start going through the textbooks to see what we need to do."

Bush soon called again to report that Stevens's chest X ray showed unusual widening of the mediastinem. "The organism is growing out profusely in his blood," the doctor told her. "He was comatose, and we never got a history from him."

After alerting the Florida State epidemiologist, Steve Wiersma, Malecki headed for the hospital. The state would have to go through the protocol of inviting CDC specialists to come to Florida for consultation, and she didn't want to lose any time. With the patient unable to speak, doctors needed his family to provide information about Stevens as quickly as possible during the weeks leading to his illness. Malecki had to know if her agency was facing a time bomb that could soon bring hundreds, perhaps thousands of additional patients into local hospitals.

She went to JFK's intensive care unit and found a private room where she could interview Maureen Stevens.

First, Malecki sketched the patient's basic biography in her notebook.

63-year-old white male; native of London.
Retired as commercial artist. Now working as photo editor at Globe Communications, which owns the National Enquirer *among other publications. He is based at their complex on Broken Sound Blvd., Boca Raton.*
At work, he mainly makes phone calls and gets photos faxed to them from all over the world. It was unknown to his wife if he receives and opens mail there (? Photos). Drives to and from work via A1A with open sunroof and music. . . . Leaves home at 8 A.M. and it takes 35–45 minutes for the drive to work.
At home she has feral cats, which she feeds. Buffo toads go after the cat food. He has relocated buffo toads by catching them in nets—taking them to lakes. Maybe as many as 50.
He is a cyclist.
Mr. S. history of heart attack 5 years ago with stent insertion; meds, 1 adult aspirin per day; Lopressor, ½ tab with breakfast and lunch. He carries nitroglycerin with him.

Malecki began taking down the details of Stevens's days, beginning on September 22:

Saturday:
1. Possibly went fishing (place unknown). His usual fishing spots include the pier at Boynton Beach Inlet (salt water, ocean fishing); Loxahatchee Wildlife Preserve (fresh water); and also Lake Osborne in Lake Worth.

He does not use a boat; he fishes from the banks. He does not hunt.
2. Out to dinner with friends at Roadhouse Grill in Boynton Beach. He had steak, a small sirloin.
Sunday:
6:20 to 8:20 A.M.. Bicycling with friend at John Prince Park, Lake Worth. Later to Lake Worth Beach with 11-year-old granddaughter to possibly the public pool or arcade there.
Monday: He worked.
Tuesday: He worked.
Wednesday: He worked late.

Maureen Stevens gave Malecki more facts about her husband's late workday. He had stayed later than usual the night before they left for North Carolina. He had been in the office alone during some of this time.

Malecki coaxed the wife to remember minute details of their trip: their gas stops along I-95, the sandwiches they ate for lunch, a description of their daughter's Charlotte apartment and her pet dog.

Malecki wrote:

She has a little puppy.
It nipped both pt.[patient] and wife.

Stevens's daughter had been ill with a sore throat for two weeks, and Maureen Stevens had been on Cipro for two weeks to treat a cough.

On Friday, September 28, the family had left for Lake Lure, near Chimney Rock, North Carolina. They toured a musty cave, then took an elevator up to a snack bar. While hiking up the mountain to an observation point, Stevens stopped and drank two handfuls of water from a waterfall.

For dinner, he had a steak, salad, and potatoes. Nothing seemed wrong.

On Saturday, they toured Charlotte and had a pub lunch—smoked salmon on brown bread. They visited the building where their daughter worked but did not go inside.

When they returned to the apartment, Stevens stayed alone while the others shopped. He walked the puppy and rested. They ate leftovers, and he went to bed before 10 P.M.

On Sunday, he stayed behind while his family went to church; then they set out for Duke University, in Durham, to visit his daughter's boyfriend. On the way, Stevens began to shiver and complained of numbness. He reclined in the car, but insisted that they continue. He rested in a Duke fraternity house while the others dined. On the way home, they passed signs for five hospitals, but he didn't want to stop.

He got up early Monday to leave for home. Maureen noticed that his pillow was soaked with perspiration. He complained of chills and wore a sweater, but insisted on driving while she navigated. They stopped twice along the way, but he stayed in the car. At home by five P.M., he sipped tea and ate a sandwich, his temperature at 101 degrees. Maureen also felt sick; she had a 102-degree fever.

Tossing fitfully in bed, he pushed his wife away and said, "Stay away from me! You're hot." Around 1:30 A.M., Maureen heard him vomiting. He was disoriented when she checked on him.

Maureen insisted on taking him to the emergency room. He walked to the car on his own, but once they arrived, he lost coherence. At 5:30 A.M., once Stevens's condition stabilized, the emergency room staff sent her home.

Maureen had returned at 8 A.M. and found him in isolation. After having a seizure, he had lost consciousness.

Throughout the reconstruction, Malecki elicited other clues that might suggest contact with anthrax. At lunchtime, he frequented Bud's Lounge in Delray Beach. He bought packaged noodles at an Oriental food store, which also sold goat meat, and liked to shop for Indian foods on Dixie Highway. He had recently purchased Borelli leather shoes.

His hobbies included collecting fishing poles and growing herbs, and he had recently weeded the garden. Mosquitoes often bit him. During two weeks in England in June, they had taken country walks and had their shoes checked for hoof-and-mouth disease at Heathrow Airport in London.

Malecki furiously took notes. A few of the items intrigued her: the late night of work in an empty office, the drink of water from the mountain waterfall, the visits to stores that might have contained skins or pelts tainted with anthrax. The next day, Maureen called to give more details about a trip the couple had made a few weeks before to feed feral cats. She had noticed a pile of rancid garbage, and Stevens had placed it in a parking lot for removal. Now, his wife wondered, could that have been the source?

On Friday, October 5, Bob Stevens passed away. The next day, Malecki went to the grieving family's home with an investigator's detachment to record notes about their final moments together.

Someone at the hospital had seen Stevens's daughter snip a lock of hair from his head as a remembrance, and this worried Malecki, since spores could have become embedded in the patient's hair.

After interviewing the daughter, the doctor's concern abated. She wrote:

i) She did not sniff the hair.
ii) She cut the hair sample from the top of his head with scissors obtained by a social worker.
iii) She placed the hair sample in a zip-lock type baggie, which the hospital social worker had also given her.

She scribbled more notes as she strolled around the Stevens property: the type of car used on the family trip, the name of the deceased's regular biking partner, the details of a neighborly dispute over trees planted along a fence. In the garage, she noticed a fertilizer spreader and wondered if it could have created an aerosol.

For the epidemiologist trying to solve a puzzle, every fact was relevant, even the family's tearful good-byes at the victim's bedside. Malecki jotted down a brief description of the last point of contact between the infected man and family members, who now must be watched as other potential anthrax victims. Before leaving the deathbed, the doctor noted, his wife hugged him, and fearing nothing, leaned over to kiss him good-bye.[44]

As Malecki completed her interviews with the family that Saturday, another team of CDC experts, headed by Sherif Zaki, was making its way from Atlanta to Palm Beach County to bring final authority to the diagnosis of anthrax in Stevens's case.

The CDC experts believed an autopsy was imperative, and after tense negotiations with the county coroner, Zaki's pathology team had received permission to autopsy the body. The examination would reveal more about the course of inhalation anthrax, which was crucial information in case more South Florida victims surfaced.

Although the Stevens family had given consent, Palm Beach County coroner Lisa M. Flannagan expressed deep reservations to Zaki about cutting open a victim of inhalation anthrax. The prospect of autopsying Stevens had disturbed her staff, concerned that the morgue would be contaminated.

True, there was a chance that anthrax spores could find their way into the morgue if blood or tissue splattered in the room and was left unattended. If left to dry out, spores might incubate under a table or in a dark, unnoticed corner. Medical advisories had routinely warned doctors that an anthrax-infected corpse was best left undisturbed.

Zaki offered to perform the procedure himself, in consultation with his CDC team, so that the federal experts would bear the greatest risk.

Perkins, then on the ground in Florida, agreed that it was essential to autopsy Stevens since doctors "were dealing with an unknown route of transmission." His one major concern was that the power saw used in a standard autopsy to provide access to the brain could be an aerosol generator, stirring up spores as it cut into the skull of an anthrax victim.[45]

Perkins and Zaki agreed to forgo the skull examination. Finally, the impasse lifted and Flanagan agreed.

During the three-hour autopsy, the CDC team wore protective suits with masks and triple layers of cut-resistant gloves.

"You always have to worry when you're in a field that's bloody. It's very easy to nick yourself on a piece of bone," Zaki said.[46]

Unlike Ebola, anthrax could not seep into a morgue technician's bloodstream through a prick in the finger. But as Zaki explained, the security precautions made sense given the messy disease that had consumed Stevens.

With a steady incision, Zaki cut open the chest, then used a ladle to clear away puddles of fluid. In a normal person, a small amount of fluid, less than two tablespoons, lubricates the area between the lung and the chest wall. But the pleural effusions of anthrax victims are high amounts of cloudy, yellow liquid, often tinged with blood.

Near the lungs, the lymph nodes, normally tan and half the size of a cherry, looked instead like ripened plums, swollen and red. Zaki removed them for study and noted the presence of hemorrhagic mediastinal lymphadenitis—enlarged, bleeding nodes around the lungs. Zaki and his team packaged the lymph tissue, samples of the fluid, and critical organs for transport back to the CDC labs.

With anthrax, the biggest biosafety risks come when procedures are finished. Great care must be taken to scrub down tables, floors, and anywhere else that anthrax-infected blood or tissue might have landed. Guidelines for anthrax autopsies suggested decontamination using chemical germicides such as iodine or carbolic acid. A hospital infection specialist whom Zaki brought along supervised the cleanup, using a strong Clorox solution to kill any lingering bacteria.

Zaki was anxious to get the tissues back to his Atlanta laboratory to try a new stain test that he had developed over the past two and a half years in preparation for an anthrax attack.

The test, similar to one Zaki had used to identify Hanta virus, involved injections with an anthrax antibody that would seek out any bacteria present and turn them bright red. This helped pathologists to determine how quickly and far a germ had spread within a body.

Zaki arrived in Atlanta with his postmortem samples around 11 P.M. on October 6 after the cleanup was completed. In the lab, his staff prepared slides to look at the composition of the lung tissue, the lymph nodes, the fluid from the chest. The pictures under the microscope left no doubt about the cause of Stevens's death. Focusing the lens, Zaki zoomed in on the final, convincing evidence. The stain test had worked. Everywhere he looked, Zaki could see tissue floating in a bath of red.

Bradley Perkins's team set out to collect anything in Stevens's daily environment that could be contaminated with anthrax spores. Perkins had called Tanja Popovic, asking her to fax or e-mail the protocols on how to collect environmental samples.[47] Stumped, Popovic did some research and realized that they were paving new ground. She decided to convene a telephonic panel of experts.

"I had about twenty people on conference call from all over the country, and we agreed. And by twelve o'clock, I put together protocols," Popovic recalled. "In other words, when you go to somebody's house, what do you do? If you wanted to see if there are spores in my office, you would take a swab of my keyboard, especially in between [keys]. You would take a swab of the corners here—take a swab of my pictures, of my floor, of my air ventilation system, of my clothes, perhaps, if I had any. So I sent him a protocol, what to do with swabs . . . how to pack them and send them to CDC."[48]

After the autopsy, Zaki brought back twenty-two environmental samples from the initial sweep of Stevens's home and workplace. About twenty CDC staffers stood by to greet him at a laboratory intake area, where they began carefully logging the samples, assigning each a number and entering it into a computer database.

The lab technicians had never seen such a random collection of stuff—samples from Stevens's yard and bike trail, clothing, carpet samples, a computer keyboard from Stevens's office desk, his entire mail cubicle from the AMI mailroom.

Popovic's staff prepared to begin around-the-clock testing. They would be working in shifts to move the process as fast as humanly possible, much of the time in the constraints of BSL3. But everyone wanted to help, calling in from home or on the road, setting up cots in every vacant cubicle. That night, a desk lamp went on in Popovic's anthrax lab that would burn continuously for the next ten weeks.

Since the lab techs could not speak to outsiders from within BSL3, they devised ways to communicate their findings. When a test result was in, a lab worker went to the laboratory window and held up an index card with the sample number and a positive or negative result. Someone logged the findings. After a string of negatives, it looked like the sampling would be a bust.

Then the lab found two positives for *B. anthracis:* the computer keyboard from Stevens's newsroom desk had tested positive, as had his office mail slot. The tests confirmed a suspicion that had been gradually building among the teams in Florida: the source of Stevens's anthrax exposure seemed to be the AMI mailroom, where letters were sorted for delivery to about a thousand other employees, all now at risk for exposure.

AMI officials had contacted the CDC on October 4 to ask if the building was safe and had been told that employees could continue working in it. That advice had placed hundreds of people in peril. Also, the response team from the CDC and Malecki's department had walked through the building unprotected, collecting samples and talking with employees about any suspicious letters or events. They all went on Cipro.

Hate mail and threatening phone calls were an occupational hazard at AMI's tabloid publications, which reveled in controversy and scandal. Earlier that year, the tabloid *National Enquirer* had been accused of libel by Carolyn Condit, the wife of California congressman Gary Condit, for its salacious coverage of the disappearance of congressional intern Chandra Levy. The tabloids only seemed to sell more papers with each expression of celebrity outrage over its slanted news coverage and surreptitious photographs.

In employee interviews, however, the response team had heard about two especially suspicious letters, one of which was the Jennifer Lopez fan letter that emitted white powder when opened. When Malecki heard about the letter, she thought it sounded suspiciously like a copycat imitating the opening scene of author Robin Cook's bestselling medical thriller *Vector.*

The response team had scoured Stevens's desk for suspicious mail but had found none. As a general rule, AMI employees sent any suspicious letter back to the mailroom, where it was held for a few days, then incinerated. The Jennifer Lopez letter could not be found.

That Sunday evening around 7 P.M., Malecki gave the order to secure the AMI building. Frightened workers who had been told all week that they had nothing to fear were evacuated and asked to line up the next morning at the health center in Delray for nasal swabs and free Cipro. The FBI's Hector Pesquera announced that a criminal investigation into Bob Stevens's death had begun, but from Washington, Attorney General John Ashcroft insisted that it was too early to conclude the case was the result of terrorism.

One other clue had surfaced that weekend, increasing the urgency of the response team's detective work. Carlos Omenaca, a doctor in North Miami, called to report that he had recently hospitalized a seriously ill seventy-three-year-old patient for community-acquired pneumonia. The patient, Ernesto Blanco, was the beloved mailroom supervisor for the *Sun.* At AMI, an alarmed executive had inventoried employees who had called in sick in the days since Stevens was diagnosed with anthrax. He knew that Blanco had been hospitalized on October 5, the day Stevens died, and the executive had persistently called the hospital until he reached Omenaca. Was it possible that the two cases were somehow related?

Blanco and Stevens were longtime work friends, although they lived more than fifty miles apart. A genial grandfather with a thick head of graying hair that made him look younger than his years, Blanco had worked beyond his retirement age because he loved his job and wanted to keep active. He had already paid off his mortgage: the work was for fun, not money. He took a commuter train to Boca Raton each day from his home in North Miami.

He had gone into work for several days in early October, though he felt strangely ill. His job involved driving an AMI van to the Boca Raton post office two miles away and picking up bags of company mail. After unloading the bags, he would dump and sort the mail, sometimes as many as five thousand pieces. Then he would deliver them by cart to various locations in the three-story, sixty-six-thousand-square-foot building. The third floor featured the offices of the *Sun,* where Bob Stevens sat. Employees were happy to see Blanco, and he often stopped to make conversation.

Blanco's boss, worried by his appearance, had been urging him to take some time off. He finally gave Blanco an order: "Go home! Get out!" He drove Blanco to the train station in North Miami where he kept his car.[49]

At home, Blanco worried that he had contracted West Nile virus. His wife drove him to the hospital, and before long he was gravely ill. His family gathered around him. A priest visited five or six times. Blanco drifted in and out of consciousness.

He did not know about Stevens's death when Omenaca showed up at his bedside and told him they needed a nasal swab to test for anthrax. The test came back positive.

Word spread that there was a second anthrax victim from AMI, and reporters from around the world flew into South Florida. The nasal swabs of 965 other AMI employees turned up one additional exposure, thirty-six-year-old office assistant Stephanie Dailey, who worked with Blanco. She reported handling a letter containing suspicious white powder. Dailey went on Cipro and soon returned to work.

Federal and local officials told the public that the anthrax release was limited to a single building and that the strain of anthrax involved in Stevens's death appeared to be a naturally occurring one. Behind the scenes, however, Popovic's laboratory had identified the strain as Ames, the one used in U.S. military labs.

For the next few weeks, Blanco struggled against the disease; a tube attached to his chest drained a seemingly endless flow of cloudy liquid. Omenaca waited until he had begun to recover to tell him about Stevens's death and about the overwhelming odds Blanco had overcome.

"You'll be all right," the doctor said. "But I thought you would never make it."

Three FBI agents came twice to his hospital bedside, grilling him about his work routine and the suspicious fan letter.

"You think that between so many thousands of letters I could pick up one and say, 'This is the one?'" Blanco asked them.

The agents were careful not to suggest that the incident had been a terrorist attack. But Blanco came to his own conclusion. "To me, it was a terrorist act. I have no doubt about it. It was nothing special against me."

CHAPTER 7

Mountain Streams and Spin

October 3–10, 2001
WASHINGTON, D.C.

Not long after Bob Stevens staggered into a Florida emergency room, HHS Secretary Tommy Thompson left his spacious office and, accompanied by a security detail, made his way across the National Mall and toward Capitol Hill. In the months since President Bush tapped him to run the behemoth social services agency, with oversight of the CDC and the National Institutes of Health, Thompson had visited Capitol Hill twice to testify on bioterrorism in sparsely attended congressional hearings. Both times, his remarks had been lost in a clamor of partisan debate and press questions about an issue then dominating the headlines, the president's position on stem-cell research.

The mood had changed and this time Thompson had been scheduled as a key speaker in a bioterrorism hearing sponsored by the Senate Appropriations Subcommittee, which oversaw his agency's $468 billion (2002) annual budget. There was no evidence that a bioterror attack was on the horizon, but since 9/11, a sense of powerlessness had sapped the American psyche. President Bush had traveled that day to Manhattan to visit a first-grade classroom near ground zero, lauding the leadership of schoolteachers who had safely evacuated 8,000 children on 9/11, as well as Mayor Rudolph Giuliani and Republican Govenor George Pataki. Thompson had done his part to help with the New York attacks by activating, for the first time, the National Disaster Medical System the morning of 9/11, with eighty disaster teams and 7,000 private-sector doctors on alert. He had stationed 570 HHS employees in New York to help with relief efforts near the World Trade Center ruins.

In the heat of the national crisis, members of Congress did what little they could to convey control. The attacks had revealed gaping holes in the federal government's security and exposed appallingly poor coordination among federal agencies charged with protecting the public. Worse yet, the blunders had come after years of warnings from government panels, independent experts, and academics that the terrorist threat was gathering and that relevant federal officials were blind to it.

Within a few weeks of 9/11, Congress had hurriedly convened dozens of congressional hearings to assess the threat and ponder the mistakes that had left America vulnerable. For federal bureaucrats, the hearings were like a call to the principal's office. Fidgeting nervously behind long wooden tables, they endured rounds of self-serving pronouncements from congressmen who often knew no more about the subject at hand than what their aides had told them in hurried briefings. Their questions often seemed designed to placate the media rather than address a pressing national problem. Yet to armchair critics watching on C-SPAN, the flurry of activity on Capitol Hill conveyed an image of a federal government that had finally sprung into action, working overtime to ensure that the agency bungling exposed by the attacks would be promptly fixed.

On the front lines of the crisis, Jeffrey Koplan thought such hearings were a needless distraction for officials already overwhelmed with the post-9/11 recovery, and he wanted nothing more than to avoid being summoned to the Hill. Still on crutches, his heavy cast replaced by a red, white, and blue bandage, Koplan stayed put in Atlanta, making few public statements without clearance from Thompson's office. The White House wanted information tightly controlled, with what little that was said carefully orchestrated.

President Bush had set the tone on October 2, emerging from a breakfast with congressional leaders declaring, "Americans also realize that in order to fight terrorism, they're going to go about their lives in a normal way. And Americans are."[50]

In reality, there was widespread speculation that America could face a second, more devastating act of terrorism, and the airwaves rang with bioterror experts predicting death tolls in the thousands should al-Qaeda decide to loose a biological agent.

Putting together a story about the possibility of such an attack, the producers of CBS's *60 Minutes* newsmagazine had contacted the White House to find an administration spokesman for a Mike Wallace interview. CBS had already booked Represenative Christopher Shays (R-Conn.), chairman of a House subcommittee on national security and a leading preparedness critic. The September 30 newscast seemed certain to add to anxiety about loosely protected chemical and nuclear plants, threatened water supplies, and vulnerable defense installations. The media handlers at the White House and HHS wanted to counter with a two-pronged message: there was no evidence that an attack was imminent, and if one occurred, America was ready.

First, they approached Koplan.

"I had done *60 Minutes* before," Koplan recalled. "The thought was, 'Maybe you could try it again. You survived your first. Try another one.'" Koplan had reservations when he was told the interview was to be scripted, with a message assuring the public that everything was under control. "I was unwilling to say that," Koplan said.[51]

The CDC director told his press office to inform HHS that he would politely decline, despite the political risk of challenging the administration. Koplan was certain that the nation's underfunded network of hospitals and clinics was in no way capable of responding to a cataclysmic bioterror event. He knew the public would see straight through the political spin.

Thompson, rapidly emerging as a Bush administration messenger of hope, had no qualms about delivering the message. He had been briefed by experts on bioterror response, and thought he could adequately face the issue, even on a news show known for its confrontational, investigative approach. When asked by Wallace about the chances of a bioterror strike, Thompson said, "It's possible, but quite doubtful." In the unlikely event of an attack, a coordinated federal response, with ample hospital beds and stockpiles of antibiotics, was ready to move on command, he said, and the Department of Defense would make additional beds available if casualties warranted it. He noted several times that there was still room for improvement, but assured viewers that there was no cause for panic.

"We've got to make sure that people understand that they're safe, and that we're prepared to take care of any contingency, any consequence that develops, or any kind of bioterrorism attack," Thompson told Wallace. "I have three kids and tonight—tonight I'm telling them that they are safe and my granddaughter, who is less than two years old, is safe as well."[52]

Wallace had confronted Thompson with frightening assessments from leaders like Shays, who ticked off the list of hostile nations believed to have biological weapons capabilities. Shays said he believed that if terrorists had access to biological weapons, he was positive they would use them. Wallace asked Thompson why there were so many doom-and-gloom reports about the nation's preparedness.

Thompson replied:

> *Mike, that is absolutely correct, and I wanted tonight, to try and explain it to the American people that I have the best doctors, researchers, scientists, individuals around me that meet daily on this particular subject. And we've looked at it, we've found that there are some shortcomings, and we have addressed those shortcomings, and we're still assessing those shortcomings.*

The *60 Minutes* appearance backfired for Thompson, who found himself the target of editorial criticism and cheap shots. The October 3 hearing of the Subcommittee on Labor, Health and Human Services and Education offered the chance to repair the damage. Since 9/11, hearings on the terrorist threat had attracted wide news coverage. Reporters who had once camped outside waiting to corner Thompson on contentious issues now crammed inside the ornate hearing rooms, listening attentively to his prepared script.

Although Bush had placed him in charge of the public health system, Thompson understood politics far better than medicine, and he had made a point to become well acquainted with the subcommittee members. With his wide, easy grin, backslapping manner, and intuitive sense of how to work a room, Thompson knew the rituals of Capitol Hill and the elaborate courting needed to sway votes and loosen the purse strings of reluctant funders. His tenure as governor of Wisconsin had made him a favorite of Republican Party conservatives, and he had been mentioned as a prospective GOP presidential candidate.

Thompson had appeared before the Appropriations Subcommittee in April 2001, listening attentively as members nominated pet programs for some of the secretary's $55 million pot of discretionary funds. Members grilled him on stem-cell research and advocated for healthier school lunches, but there was little mention of bioterrorism preparedness.

By October 3, priorities had changed. Congress seemed inclined to throw money at any proposal that might avert another disaster. Two senators, Bill Frist, a Republican from Tennessee, and Ted Kennedy, the Massachusetts Democrat, had pushed to more than double federal spending on public health preparedness for bioterrorism. Frist believed a germ attack was inevitable, given Osama bin Laden's pronouncements about acquiring biological and chemical weapons. "The threat is real; there is no question about that," he said in a television interview that day.[53]

Thompson, in contrast, told senators he was heartened by HHS's bioterrorism preparedness. Some improvements were still needed, but the public health surveillance network had been upgraded to improve detection of the release of biological agents. Antibiotics and vaccines had been stockpiled, and government regulation had stymied the dangerous transfer of biological agents. Thompson said:

> *We had two major cities hit simultaneously with terrorist attacks. When we began to respond that morning, we didn't know if there was bioterrorism involved, and we didn't know how many injuries or casualties there would be. Yet, we immediately implemented our health alert system at the Centers for Disease Control and Prevention. . . . This is the first time our emergency response system had been tested at this extreme level, and it*

responded without a hitch. . . . And it should encourage the American public that we do have the ability to respond.[54]

Senator Robert Byrd, a Democrat from West Virginia, skeptically questioned Thompson about his comments on *60 Minutes.*

BYRD: In a recent media interview, you said that the U.S. government is "prepared to take care of any contingency, any consequence, that develops from any kind of bioterrorism attack." That's a pretty broad statement.

THOMPSON: It is.

BYRD: Do you stand by that today?

THOMPSON: I do. I'd like to clarify. I said we could respond. Evidenced by what we did on September 11th, I am absolutely assured that we can respond to any contingency and control it.

BYRD: That's a broad statement. And Washington is so full of hyperbole and broad statements, we know, we should know because we make them too. I tell you, that's a bad thing to do, if we mislead, and I know you don't intend to. I know you don't intend to.

THOMPSON: I'm telling the American people so that people understand that we are prepared.

BYRD: Well, I just don't believe that. And I say that to you very kindly.[55]

In his prepared testimony, Frist offered the subcommittee a harshly different perspective from Thompson's, describing a bleak scenario of overloaded hospitals, crashed communications networks, and general pandemonium in the event of a biological release. A year earlier, Frist and Kennedy had pushed through the first bioterrorism preparedness legislation, the Public Health Threats and Emergencies Act, allocating $540 million in federal monies for hundreds of hospitals and health clinics to help improve their emergency response. The senators now proposed increasing the funding to $1.2 billion.

"We have allowed our public health system—the front line of our defense—to deteriorate over the past twenty years," Frist said. "We must buttress our local response by upgrading local and state medical surveillance epidemiology, assuring adequate staffing and training of health professionals to diagnose and care for victims of bioterrorism, and improving our public health laboratories, many of which simply are not equipped to efficiently diagnose infections and other diseases associated with biochemical weapons."[56]

A surgeon who had left Vanderbilt hospital's operating room to run for the Senate in 1993, Frist was the Senate's only physician, a statuesque, square-jawed man set apart by his youthful looks and manicured dark brown hair. He had quickly won favor with the Bush administration, which would support him in a late 2002 bid to take over the job of Senate Majority Leader from Louisiana's embattled Trent Lott. In 1997, he had held hearings to assess the medical system's ability to respond to mass casualties and had kept up the drumbeat until the preparedness act passed with bipartisan support in 2000. Yet Frist remained convinced that a bioterrorism incident was inevitable and that federal readiness efforts lagged far behind need.

Part of Frist's determination was fueled by his own experience with bioterror. A few years earlier, a suspicious letter had landed in the office of a public health subcommittee he chaired. The letter was tucked inside a yellow envelope with the word *anthrax* scrawled upon it. Frist's staff turned the letter over to authorities. Like dozens of others that surfaced during that time, it proved to be a hoax, but the incident convinced Frist that he and others on Capitol Hill were especially vulnerable.

Public officials were likely targets because of their visibility. "You pick a high visibility person," he said, "because then terror can be personalized."[57]

Even after the passage of his bill, Frist spoke constantly about preparedness, often seeming like a solitary voice, preaching into a vacuum. But when the news of Stevens's illness hit Capitol Hill, Frist suddenly seemed like a prophet. It became clear to him that "no longer was bioterrorism a risk on our soil—it was a reality."

Despite his feel-good public statements, Thompson also had become more concerned about a bioterror attack. On September 16, he had hired as an adviser Dr. D. A. Henderson, head of the Johns Hopkins University biodefense center. The critical and sometimes eccentric Henderson was dissatisfied with U.S. bioterror preparedness, and his institute at Hopkins had pinpointed inadequacies with its elaborately staged Dark Winter exercise that summer. In 1998, Henderson wrote, "The specter of biological weapons use is an ugly one, every bit as grim and foreboding as that of a nuclear winter. As was done in response to the nuclear threat, the medical community should educate the public and policy makers about the threat."[58]

By adding Henderson to his staff, Thompson signaled his willingness to reexamine the problems.

At the same time, he had been pressing President Bush in private to approve a $1.6 billion bioterrorism package that included an ambitious plan to stockpile 300 million doses of smallpox vaccine, enough to inoculate every American against another potential terror agent. Thompson's efforts finally bore fruit. Bush gave him a verbal commitment in late September, and in early October the secretary announced that HHS had arranged with its principal supplier to speed up smallpox vaccine production and make it available by the summer of 2003.

Mobilizing a speedy vaccine program response to the anthrax attacks was not so straightforward. Mired in controversy, mass production of the U.S. anthrax vaccine had been stalled.

On October 4, as CDC lab cultures confirmed that the disease infecting Stevens was anthrax, Koplan called to pass along the alarming news to Thompson, who then headed for Pennsylvania Avenue to inform President Bush. Some Bush aides said later that when the president was informed of Stevens's death, they saw fear in his eyes for the first time since 9/11. Thompson said the president showed concern, not fear, then turned to him and said, "You go out and tell the press, Tommy." Bush told Thompson the message had to go out to the public as soon as possible.[59]

Through a coincidence of timing, Bush's press secretary, Ari Fleischer, was heading down the hallway to give his daily briefing to the White House press corps, the most aggressive political reporters in Washington. Amid shouted queries from reporters jockeying for attention, the press secretary used these daily sessions to review major national and international developments, occasionally introducing a key administration figure to take questions about a breaking news story. Most were accustomed to gentler treatment. Press corps pros grilled relentlessly, pushing for the sound bite that might make the evening news or the morning headlines. The spectacle sometimes called to mind the biblical story of Daniel cast into the lions' den.

Having survived the Mike Wallace interview, Thompson stepped forward at the afternoon briefing to announce that the CDC had confirmed Stevens's diagnosis. "Based on what we know at this point, it appears it is an isolated case," he said. "I want to make sure that everybody understands that anthrax is not contagious and is not communicable, which means it does not spread from person to person. If it is caught early enough, it can be prevented and treated with antibiotics."

Thompson said the CDC was investigating Stevens's movements to try to find a natural cause for the illness, reminding the public that anthrax does occur naturally in the United States. Nevertheless, he said, the public health system was on heightened alert.[60]

Then the media questions began. A reporter asked if the Florida patient had been in contact with wool or other materials traditionally associated with anthrax.

THOMPSON: That's entirely possible. We do know that he drank water out of a stream when he was traveling to North Carolina last week. But as far as wool or other things, it's entirely possible.

Q: Can you explain why he was drinking from a stream? (Laughter.) Is that a reason—should we know that? Why are you giving us that detail?

THOMPSON: Just because he was an outdoorsman, and there is a possibility that—there are all kinds of possibilities.

Q: Can you contract it that way, is why she's asking—can you contract anthrax by drinking unboiled water?

THOMPSON: We don't know yet.

The news conference left the impression that Thompson was either woefully uninformed or deliberately downplaying the severity of the case, making reporters even more skeptical of his statements. In the science and public health fields, the secretary's comments were regarded as laughably naive. Thompson said later it was a "mistake" to pass along information from a CDC memo about Stevens's sips from the mountain stream.

"None of us really thought that what Secretary Thompson said about the person getting anthrax from hiking was true," remembered New Jersey State epidemiologist Eddy Bresnitz. "We hadn't had a case of inhalation anthrax in twenty-five or thirty years in this country. Having it occur in a person working in a media building didn't make any sense whatsoever. And I don't think anybody at the CDC believed it either."[61]

At USAMRIID, John Ezzell heard about Thompson's comments, and his instincts told him the case must be far more serious than the secretary had portrayed. The chances of contracting anthrax through water were practically nonexistent. Ezzell called his daughter, then working at Orlando's Disney World, and urged her to go home to North Carolina.

At the CDC, Koplan, still heeding the White House directive, hesitated to comment publicly. He gave an interview suggesting that the Stevens case might have come to national attention because of an improved detection system. "What might have been tossed off as an undetermined bacterium was sent on to a state lab, where people recently received training in detecting anthrax," he said.[62]

Koplan disagreed with criticism of Thompson quickly surfacing around the country. The secretary, he said, had been acting on information collected by the CDC team, the best available at the time.

"We had people in North Carolina. . . . We learned that [Stevens] rode a bike around on a path in Florida. We were asking the folks down there, 'Check the path. What houses did he go by?' It could have been anything—from a bunch of terrorists with a lab in a house on the outskirts of this bike route.

"Poor Secretary Thompson was given this as background," Koplan recalled "And I don't think he pretends to be a physician, a scientist, or an anthrax expert. The further removed you are from an investigation the weaker you are in your ability to answer questions and interpret information."[63]

The smooth interagency coordination that Thompson had described in his *60 Minutes* interview fell apart instantly when state and federal officials realized they had a desperately ill victim of inhalation anthrax on their hands. In Palm Beach County, confusion reigned as authorities from a multitude of federal, state, and local government agencies scrambled to decide who was in charge and just how much the public needed to know.

Once the anthrax case was established as criminal in nature, the FBI moved from a secondary to a primary role in managing the investigation. The FBI's Miami field office took command, reporting back to the weapons-of-mass-destruction unit at Washington headquarters, headed by agent James Jarboe. In Florida, early conflicts between the CDC's epidemiological team and the FBI criminal investigators over evidence collection and witness interviews made it obvious that two strikingly different cultures were at work.

The health team, led by Perkins, was looking for patterns that might help it develop a working hypothesis about the disease and how it had been contracted. The FBI wanted to find the perpetrator. Occasionally, these missions collided head-on. Perkins would call Koplan, who in turn would call Thompson at HHS. Thompson had his own daily briefings at the White House, attended by Attorney General Ashcroft, and it was there that many issues of jurisdiction and procedure were discreetly worked out.

The FBI's posture was to "roll out the tanks," Perkins later recalled.[64] Every slip of paper was evidence; every human a potential witness to take before a skeptical jury. For most of the federal crimebusters, the medical and scientific aspects of the case were unfamiliar ground.

To defuse the tension, Perkins tried cross-pollination. He allowed an FBI agent to join the inner circle of his epidemiological team, while one of his CDC disease specialists joined the FBI squad as a consultant. The CDC also dispatched Dr. Mitch Cohen, the director of its Division of Bacterial and Mycotic Diseases, to Washington to serve as a liaison to the FBI. Cohen, who had trained in both medicine and microbiology, had the technical expertise to advise the bureau on medical aspects of inhalation anthrax and such scientific issues as the nature and composition of anthrax spores. FBI officials told Cohen they welcomed the help since the bureau's ranks contained few microbiologists and physicians. Jarboe's unit helped Cohen set up in a cubicle at the Hoover Building headquarters, and Cohen stayed for three months, trying to bridge the culture gap.

Each day, Cohen would consult with a team of up to ten agents gathered in an FBI conference room. He would inform the agents of the latest medical findings, then explain what they meant, sometimes illustrating points by sketching simple diagrams. He compiled a "line list," a detailed listing of patients and their identifiers—and described the type of evidence the CDC needed to declare a case "confirmed" rather than "suspected." When Popovic's lab determined that the anthrax in the Florida samples was the Ames strain, Cohen explained to the FBI that this finding, while significant, did not necessarily pinpoint a laboratory as its source because of frequent swapping among scientists.[65]

At the local and state levels, Malecki and Florida epidemiologist Wiersma wanted to get as much accurate information to the public as they could, including the sensitive determination that the material was the Ames strain. After a call from Florida governor Jeb Bush's office, citing national security interests, they deferred to political higher-ups.

The better part of every day was spent in conference calls. Koplan each morning started with a 9 A.M. briefing from his doctors in the field, but the briefings often overlapped with conference calls initiated by HHS, the White House, or the National Security Council. Sometimes, as a conference call was almost over, a harried official would run into the room with new information, shouting that the picture had changed suddenly. Koplan's CDC command center became a well-greased twenty-four-hour operation. In a converted auditorium in Washington, outside Thompson's office, a similar center was going full tilt. Televisions blared the latest news from CNN while workers punched every new scrap of information into a bank of humming computers.

Each morning and afternoon, and more frequently if problems surfaced, representatives of the FBI, CIA, CDC, HHS, USAMRIID, and other federal agencies shared information and sorted out their problems in top-secret conference calls coordinated by the NSC's domestic terrorism expert Lisa Gordon-Hagerty. Hagerty reported back to Tom Ridge at the White House, who was still setting up his homeland security office when the crisis began. The CDC usually took the lead, updating other agencies with details from Florida. Dr. Julie Gerberding, the acting deputy director of the National Center for Infectious Diseases and a regular participant in the calls, considered the sessions "the most detailed and probably the most effective intra-agency coordination."[66] At first, she wished she could get more information about the FBI's progress, she said, but she reconciled herself to the fact that the Bureau was operating under different rules.

At that point, the FBI actually knew very little. Attorney General Ashcroft told reporters on October 8 that even with the discovery of a second Florida anthrax case, the sick AMI mailroom supervisor Ernesto Blanco, there was still not enough evidence to deem the matter a terrorist investigation. Asked if hijacker Mohammed Atta could be connected, Ashcroft said, "We haven't ruled out anything at this time."[67]

Around the country, FBI agents chased thousands of bogus leads and quieted outbreaks of public concern. In eighteeen days in early October, it handled 3,300 calls of threats related to bioterrorism, including 2,500 involving anthrax.[68]

In Hialeah, Florida, agents rushed to a home whose occupants had reported opening a manila folder filled with powder, then cordoned off the block and donned protective gear to inspect the premises. In Milwaukee, agents armed with a search warrant raided the home of an alcoholic scientist, twice fired from his job at Battelle Memorial Laboratories, a government contractor, after he reported making anthrax in his basement. In Ohio, agents paid yet another visit to Larry Wayne Harris to check on the workings of his home laboratory.

"I want you to know our investigators are hard at work . . . responding swiftly and fully to each and every request for assistance, but most particularly following up on each and every lead that could disclose the identity and provide the proof against those who are responsible for these anthrax attacks," Ashcroft said.

Over time, Thompson and the White House realized they were only compounding the problem by silencing authorities who might help the public understand the scientific and medical aspects of an unprecedented event. They began allowing CDC doctors to speak directly to the public, leading to almost daily media teleconferences with anthrax specialists like Perkins and a full schedule of television appearances and media interviews with Koplan and other CDC officials.

But the damage had been done. Abigail Salyers, a past president of the American Society for Microbiology, remembered the clamor of phone calls from microbiologists seeking guidance and information to give to the public about anthrax—not the sensitive facts of a criminal inquiry, just details about the symptoms of the disease and the nature of spores. There was a pervasive feeling among scientists, she said, "that the government wasn't doing anything."

She expected the CDC to mobilize a vast public information network, using the nationwide university system to disseminate vital facts. Instead, word spread throughout the scientific community that the CDC was under a White House gag order and could not respond to simple inquiries.

Salyers remembered thinking at the time that if the anthrax that killed Bob Stevens was released as a terrorist act, the White House had helped the culprits achieve their mission. The freeze on information had only added to the public perception that events were spiraling out of control. At that point in the case, the cause of much of the public's fear was not the terrorists. "It's the U.S. government," she told herself.[69]

CHAPTER 8

Media Madness

October 12, 2001
NEW YORK CITY

Journalist Johanna Huden had hardly had a moment to catch her breath since she climbed to her apartment rooftop on 9/11 just in time to watch the second jetliner smash into the World Trade Center, the same image Leroy Richmond watched in silence in Brentwood's cafeteria. With Lower Manhattan smoldering, covered in a blanket of white ash, the attack had been the beginning of an exhausting week at the feisty tabloid the *New York Post*. Normally, Huden, thirty-one, compiled the *Post*'s entertainment calendar, sorting through a daily avalanche of mail to find items worth listing. But in the frenzy after 9/11, with all hands assigned to a single mission, an editor had ordered the three-year *Post* veteran to write an eyewitness account of the rooftop scene, then to interview the mother of a pilot whose hijacked jetliner crashed into a Pennsylvania field. She would stumble home late at night and crawl into bed, and just as she was dropping off to sleep, two noisy art students living upstairs would rev into action, ruining her chances of rest.[70]

Then, there was the strange blister on her long, thin right middle finger. She had first noticed it on September 21 while attending a friend's wedding, but had tried to ignore it, even when it itched and oozed like an infected pimple. She had always had blotchy skin given to flare-ups, and lately she had been working herself to near collapse. Maybe the frenetic lifestyle of the big-city journalist, usually fueled by cigarettes, stiff cocktails, and even stiffer coffee, had caught up with her.

Days passed before she showed the wound to her boyfriend, who gasped when he saw it and ordered her to go to a doctor immediately.

The doctors diagnosed an infected spider bite and, after searching in vain for the stinger, put her on the antibiotic Augmentin. But the sore kept getting worse. She went to two clinics and saw six doctors, with no success. Finally, alarmed by the sore's appearance, her boss demanded that she go to a hospital. On October 1, Huden went in for surgery at New York University Medical Center.

Huden's surgeon cut the wound open, then sliced off and discarded the dead skin, not considering that the blackened tissue merited further study.

"They sent me home, and it's right on the knuckle, so I can't bend my finger," she recalled. "I go to the hand surgeon the next day, and it's just disgusting. That night, when I had to change it myself I almost fell over. It wasn't oozing—it was the tissue under there, you could see the yellowy tissue under there."

Huden, facing an incredibly busy schedule, tried to push the injury out of her mind. She had to visit the doctor three times a week and was wearing a brace on the injured finger. Back at work, there was a new national scare, the anthrax case down in Florida. A photo editor had died from it, and the FBI seemed to think it was somehow connected to a letter.

Huden was in the office on October 12 when a report broke that Erin O'Connor, thirty-eight, an editorial assistant to NBC news anchor Tom Brokaw, had been diagnosed with cutaneous anthrax. Her coworkers gathered around the television, and Huden listened to the details with mounting concern. O'Connor had the least severe form of anthrax disease, but it had almost certainly been caused by a deliberate act of malice. The report suggested that the case could be linked to a suspicious letter that had passed through NBC headquarters in Rockefeller Center.

Huden walked back to her desk, sat down at the computer, and typed "cutaneous anthrax" into her Internet browser. A list of items popped up. She clicked on one of the listings, which called up a photograph of a black-scabbed anthrax lesion.

"Oh, God," she thought to herself amid the din of the suddenly energized newsroom, with editors strategizing about how to pounce on the story. "That's what I had!"

At around 3 A.M. on the morning of October 12, a groggy Jeff Koplan flashed his security card at an electronic sensor and pushed through the doorway leading to Sherif Zaki's pathology laboratory. He had pulled himself out of bed and driven back to the CDC at this early hour for a single reason: to look through Zaki's microscope. What he would see there might escalate the anthrax case from a serious but limited Florida incident to a much broader case that could signal the beginnings of a national emergency.

Zaki's staff had prepared slides made from a biopsy of Erin O'Connor's lesion. O'Connor had been taking antibiotics since October 1, when she went to a doctor complaining of a mild fever and skin rash, and was recovering. After the Florida cases surfaced, and information about the disease saturated the news, her doctor alerted New York City health authorities that he was treating an unusual wound that might be an anthrax lesion. There was little urgency, just concern that her situation resembled that of the Florida journalist. Both patients worked in media operations, and O'Connor handled Brokaw's mail.

The New York City Health Department did the initial analysis, and Dr. Marcie Layton, who ran the communicable disease division, called the CDC's Stephen Ostroff, who had helped New York grapple with a an outbreak of West Nile virus two years earlier. Layton told him that preliminary analysis of the biopsy suggested cutaneous anthrax. But city officials were wary of the tests. O'Connor had guided authorities to a September 25 letter, postmarked St. Petersburg, Florida, which she believed could have brought her in contact with the deadly bacteria. She remembered finding white powder in the envelope and scratching herself after handling the letter in the same spot on her chest where an ulcer later appeared. Tests of the letter and envelope, however, had found no anthrax. The city wanted expedited laboratory work from the CDC to sort out the mystery.

Ostroff agreed to help, and within hours, the sliver of O'Connor's skin, embedded in paraffin, had been dispatched on a flight to Atlanta. Zaki geared up his staff of six. Normally, the tests would take eight hours, but with expedited treatment, they could have a reading in as little as four hours. Since O'Connor had been taking antibiotics for eleven days, the results likely would not be definitive. Zaki knew that the drugs would have killed off some of the anthrax bacteria, and perhaps none would even be visible when he administered his stain test. "It was really hanging on this one test . . . and there were a lot of people who would rightfully question how specific, how good a test, it is," Zaki said.[71]

Zaki felt that his reputation was on the line. If he failed to spot a case of anthrax, authorities might postpone tracking down the disease and finding other possible victims. If he judged the sample to contain the bacteria, he knew that within hours pandemonium would erupt.

Koplan did not want Zaki to face the consequences alone. He and Jim Hughes from the National Center for Infectious Diseases told Zaki they wanted to be there with him when the results were in, no matter what the time. Zaki had worked often with his boss Hughes, showing him important laboratory results. But Koplan had never before personally reviewed his findings.

"It was very important to feel the trust," Zaki said. "If you're on your own and don't have the backing of your superiors, that's very difficult. It helped make the final decision. Sometimes you can't be a hundred percent sure. . . ."

The three men took turns examining the tissue. Fragments of red stood out against a white background. The antibiotics had been working, but the stain test showed they had not yet totally wiped out the germ.

The officials concurred. The sample was positive for *B. anthracis*.

Before daybreak, Koplan walked on crutches through the CDC's labyrinth of hallways and sat down in his office to call Layton and tell her the news. New York had been so overwhelmed by 9/11 that

Koplan could not imagine city health officials taking on another burden, especially a potential case of bioterrorism. He also feared that the case had broader significance. If it had been a terrorist act, the terrorist's reach had gone beyond one state to a second, and perhaps to others.

Koplan broke the news to Layton, and a few hours later, his phone rang with a call from City Health Commissioner Neil Cohen. Cohen asked him to hold for Mayor Rudolph Giuliani.

"Does she have it?" the mayor said.

Koplan began offering his opinion, filled with the usual medical qualifiers.

The mayor interrupted. He had no time for subtleties. "Does she have it or not?" he asked impatiently.

Koplan sighed. "Yes, she has it."[72]

Having witnessed the New York media frenzy during the West Nile scare, Stephen Ostroff knew the drill: he threw together some clothes and boarded an executive jet bound for LaGuardia Airport. A police escort whisked him to the Emergency Operations Center at Manhattan's Pier 92, where Giuliani and Pataki were waiting. Although the September 25 letter had shown no traces of bacteria, a team from the CDC and the city's health and police departments was ready to scour NBC's offices at Rockefeller Center for any clues, using the same methods employed in the Stevens case. Ostroff knew that only by surveying O'Connor's workspace could the source of her infection be determined.

That morning, Cohen's department announced to the public that an unidentified NBC employee had been diagnosed with cutaneous anthrax, citing the September 25 letter as the possible source. NBC president Andy Lack and chairman Bob Wright sent a memo to the network's employees, stressing that the skin infection was not the same respiratory anthrax found in Florida. From Washington, HHS Secretary Thompson said there was no proof that the case was a terrorist event. And the letter's intended target, Brokaw, said in his nightly newscast that the incident was "so unfair and so outrageous and so maddening, it's beyond my ability to express it in socially acceptable terms."[73]

Giuliani announced his intention to close one floor and some additional work areas in NBC's building so that samples could be taken, but noted that the step had been ordered out of an excess of caution. There was no plan to close down the entire building. Ostroff, who stood with him to face reporters, remembered being battered with questions about why the September 25 envelope tested negative. "And I had the foresight to say, 'Well, maybe it just wasn't randomly distributed throughout the envelope. There wasn't much material there, or maybe the tests aren't that good, or maybe it's not the right letter.'"[74] Ostroff suspected that the cause of O'Connor's infection might still be lurking somewhere near Brokaw's office.

NBC's building is an Art Deco treasure constructed in 1932, and includes a shopping concourse and scores of corporate offices. Crowds gather daily between busy Fifth and Sixth Avenues, hoping to get a glimpse of NBC's star newscasters or to see their image pop up on the screen of the morning *Today* show.

Into this scene came workers in hazardous materials garb, ready to survey the third floor, and health department workers intent upon doing nasal swabs on thirteen hundred employees and distributing Cipro to those who wanted it. Allan Maraynes, a senior investigative reporter for NBC's *Dateline,* recalled listening with growing skepticism to reassurances from company executives and health officials that the fourth-floor office where he worked was a safe distance from the third-floor "hot spot."

"We walked in one morning, and we're told that some interesting mail had been received, and it is reported to have contained a strain of anthrax. At that point, everybody broke into a mild sweat, looked around, and wondered if they should freeze or head for the exits. Nobody panicked, and we were told there was no need for panic," he said. "There was still a normal question mark. Should we be concerned? Do we believe them? There were people here from CDC—guys in moon suits. We were walking around without moon suits. . . . It was disconcerting."[75]

From her home, O'Connor helped guide the search, telling the FBI about an interoffice envelope that she always kept near her desk containing a collection of suspicious mail. An NBC security guard hurried down from the sixteenth floor to retrieve it, then carried it in a plastic grocery-store bag back to his office, where he waited for officers from the Joint Terrorism Task Force to take it away. Inside was a letter postmarked September 18 from Trenton, New Jersey, with the envelope stapled to the back of the page.

When the officers arrived, Ostroff remembered, they carried the plastic bag to a second floor mailroom, opened and inspected the letter, then took the materials out of the building to the city laboratory. The letter tested positive for anthrax, and Ostroff's team had its source. But the letter's routine path through NBC's building had left a trail of contamination.

"Every place along that route basically turned out to be positive. The sixteenth floor, the third floor, the second floor—there was even a spore in front of the elevator shaft," Ostroff said. "And then the mailroom, of course, got contaminated. And then when they opened it in the city laboratory, the city laboratory got contaminated, and there was essentially almost no material left." Two police officers and a lab technician tested positive for anthrax exposure.

Despite the fact that the contaminated letter had been carried through other parts of Rockefeller Center, Ostroff saw no reason to close the entire building based on his understanding of anthrax fundamentals. Once it settled into an environment, anthrax powder was believed to be stable. Experts did not believe it could re-aerosolize, at least not "in a quantity that is sufficient to produce subsequent disease," Ostroff thought. At NBC, the letter had been inside the building for several weeks, and thus far, no one else had gotten sick from it. "Why would we close down a sixty-five-story office building with all the disruption associated with that for what I couldn't discern as being a significant health risk at that time?" he asked.

The trail of contamination presented other problems. The city lab was shut down for several days for cleanup. When it reopened, the lab was "absolutely flooded with every powder, everything you can imagine from the city of New York. Tabletops and mirrors—you name it, they had it—and it was just backing up tremendously in the laboratory," Ostroff said.

The CDC team used its Laboratory Response Network to find other labs capable of doing the testing. "We did manage to send a bunch of specimens up to Albany. . . . We ended up going to the

Department of Defense, which has rapid sampling and testing methodologies that they use on the military side. After a great deal of discussion with the Pentagon, we managed to get one of the teams to come up to New York, Tuesday or Wednesday of that week," Ostroff said.

Everywhere, news of the New York case had heightened awareness, and laboratories were flooded with specimens. In Nevada, state health officials were holding a letter sent to a Microsoft licensing office in Reno, postmarked from Malaysia, which was known to have been a meeting spot for al-Qaeda terrorists. The initial test on the material was positive for anthrax, but further testing showed that the powder was benign. The FBI kept details about the letter's markings under wraps, their suspicions growing that some of the hoax letters could be linked to the same perpetrator.

A few blocks away from NBC, police on October 12 had cordoned off Forty-third Street between Eighth Avenue and Broadway, where the *New York Times* newsroom was tested after veteran reporter Judith Miller received a threatening letter tainted with talcum-like powder. Miller had coauthored *Germs,* a nonfiction bestseller about agents of biological warfare. Her letter also proved to be a hoax.

The case involving the seven-month-old surfaced on October 12, when NYU Medical Center's chief of pediatric infectious diseases, Dr. William Borkowsky, reported a hospitalized infant boy with a lesion on his left arm. The child's mother was an assistant to ABC anchor Peter Jennings. Dr. Borkowsky had also heard about a case at NBC and about Huden's possible case at the *New York Post.* All of the patients had seen clinicians who suspected infected bug bites.

But of all these reported cases, the baby's illness was the most serious. A baby-sitter had brought the child into the ABC building to spend some time with his mother, and the baby had spent ninety minutes inside her office and the cafeteria. The mother took him with her to a birthday celebration for a colleague on another floor, where he was held by a few of his mother's admiring coworkers.

A day later, the baby's arm ballooned and his mother noticed a small red lesion, the size of a half-dollar, on the back of his arm. It oozed yellow liquid and was hot to the touch. A doctor placed the baby on antibiotics, but about a week later, he fell deathly ill and was rushed to NYU's pediatric intensive care unit. By then, the lesion had formed a black scab.

Anthrax toxins had surged into the baby's system and were rapidly destroying his red blood cells. His kidneys were failing and doctors diagnosed disseminated intravascular coagulation, which basically meant he was at risk of bleeding to death.

After Ostroff's team heard about the child, a hunt for a suspicious letter began at ABC, but no letter was found. Tests showed a positive anthrax reading on a slot in the office mailroom. Two days after the ABC case surfaced, Ostroff's team heard about a letter at CBS and moved to its headquarters on Fifty-seventh Street, roping off an area near anchorman Dan Rather's office. On October 18, authorities confirmed a cutaneous case of a CBS employee, Claire Fletcher, twenty-seven.

Finally, the team decided simply to test every major media outlet in the city. Most turned out negative. But at the *New York Post,* an e-mail went out to employees asking them to look through their work-stations for suspicious mail and to throw all of it into a bin for inspection. The FBI checked the bins and soon discovered the Trenton-postmarked letter that had infected Johanna Huden.

"The lady knew she had a lesion, but it was thought she had a bug bite," Ostroff said. "If you look epidemiologically, she's the first case in New York. The thing that was unusual about that letter was that it wasn't opened. It was taped shut. In talking to people in that area, none of them recalled ever having seen it. And you could never quite tell who threw it in the bin," he said.

Once the letter was found, a number of people came to inspect the area and three of them developed skin anthrax. Others were vulnerable, Ostroff said, because "the mailbag that they put all the letters into as they were going from floor to floor had a hole in the bottom of it, so it was trailing anthrax spores all over the building. Even though the letter was unopened, there were a lot of anthrax spores coming through the pores of the letter."

When the letter was discovered, Huden and a group of coworkers rushed to Mt. Sinai Hospital for testing. The nurse told them it was unlikely they had anthrax, but Huden told her she had recently been treated for a strange black lesion. The nurse moved her to the front of the line.

The first test of a biopsy from Huden's wound came back negative, but a second blood test showed positive results.

The city health department put out a release announcing that a probable incidence of cutaneous anthrax had been found and warning that other cases related to the *New York Post* letter might surface.

But by then, Ostroff had returned to Atlanta to deal with larger problems. Cutaneous anthrax had shown up in another victim, a New Jersey mail carrier. And still another anthrax-loaded letter had surfaced bearing a Trenton, New Jersey, postmark. This time, the recipient was in Washington, D.C. The intended victim was the majority leader of the United States Senate.

CHAPTER 9

The Face of Satan

October 15, 2001
FORT DETRICK, MARYLAND

John Ezzell stood by the guarded side entrance to USAMRIID waiting for the FBI to arrive with the evidence, as he had done so many times before during anthrax scares. Ezzell and his team knew that with this package, coming straight from Capitol Hill, the eyes of the world would focus on his Special Pathogens Sample Test Laboratory. They did not consider its danger until later, when Ezzell opened the envelope to confront what he could only describe as "the face of Satan." Out burst a spore powder so pure that it evaporated in midair.

The intern in Senator Daschle's office suite had cut open the taped business envelope around 9:45 that Monday morning, October 15, and it let out a puff of airy white powder. Five minutes later, the U.S. Capitol police sped to the scene on the sixth floor of the nine-story Hart Senate Office Building. Luckily, Daschle was out of the office at the time, but about forty of his staffers were on duty. Someone had shut down the ventilation system around 10:30 A.M. to stop the powder from wafting through the building, but the material had already become airborne within the office, potentially endangering hundreds of staffers, as well as the daily parade of lobbyists, federal workers, and tourists passing through the Senate building.

The Capitol police did preliminary rapid swabs of the material, estimated at about two grams in quantity, the rough equivalent of one-half teaspoonful. It tested positive for anthrax. Authorities examining the letter picked up on disturbing similarities between the Daschle letter and those that had infested New York media outlets with anthrax. The letters sent to NBC and the *New York Post* had read:

This is next
Take Penacilin [sic] Now
Death to America
Death to Israel
Allah is great

The Daschle letter was worded differently, but was written in the same crude hand. At the top it was dated 9/11/01. The text read:

You can not stop us
We have this anthrax
You die now
Are you afraid?
Death to America

The Daschle letter's envelope was the first to include a return address. Written as if by a schoolchild, it read: "4th Grade, Greendale School, Franklin Park, N.J. 08852."

Chaos erupted on Capitol Hill. Mail deliveries were suspended, and bundles of letters and packages were quarantined. Tours of the Capitol were canceled. Health-care workers arrived on the scene, led by the Capitol's own attending physician, Dr. John Eisold, and followed by doctors from the NIH, the CDC, the U.S. Surgeon General's Office, and the D.C. Department of Health. With lightning speed, antibiotics arrived and staffers fearing exposure lined up to get their supplies. An outraged Daschle told reporters that he had sent word to other congressional leaders that they, too, should be on guard. Soon spores turned up in locations scattered around the

Capitol office buildings, prompting discussions about shutting down the whole complex.

Targeting the Hart Building may have carried more symbolism than its perpetrator intended. The building, which opened in 1982, had once been tagged by rabble-rouser Senator William Proxmire (D-Wis.) as a symbol of self-serving government waste, a Taj Mahal on the Potomac. It took ten years and $137 million to build (making it at that point the most expensive federal building ever constructed), and by the time it was finished, the structure stood as such a testament to government excess that no senators wanted to move into it. The building, which abutted a neighborhood of charming shops and historic row houses, featured solid brass elevator doors, offices with sixteen-foot ceilings, and expensive teak paneling. Tourists flocked to behold the vast atrium, made of Tennessee marble, and its centerpiece, a five-story Alexander Calder sculpture called *Mountains and Clouds,* with black aluminum clouds suspended in midair over black steel mountains. An inscription over the building's entrance seemed unintentionally ironic. It honored former senator Phillip Hart (D-Mich.), whose "humility and ethics earned him his place as the conscience of the Senate."

Once the controversy over its cost died down, the posh quarters, which featured office suites for fifty members as well as several important committees, became more desirable. Daschle's staff worked from a two-story suite centered on the fifth floor, beside one occupied by Senator Russell Feingold (D-Wis.). Each workday, the office buzzed with efforts to service the needs of Daschle's constituents in South Dakota. The office grunt work, including opening and processing bundles of daily mail, fell to a small crew of interns—ambitious college students willing to work long hours at low pay to learn the ropes of Congress.

A former Air-Force intelligence officer elected to the Senate in 1986 after eight years as a congressman, Daschle had recently assumed a position of much greater prominence. When Vermont Republican Senator Jim Jeffords declared his independence from the Republican Party in May 2001, he broke a delicate balance of power in the Senate and abruptly shifted control to the Democrats. Daschle subsequently took over the majority leader's post from Senator Trent

Lott (R-La.), who stirred up acrimony by suggesting that Democrats may have won a plurality but lacked what he called the "moral authority of the voters" needed to lead the chamber. Senate chairmanships shifted to the Democratic side, often bringing major changes in the philosophical bent of the major committees.

Much of the summer had been spent trying to construct a bipartisan compromise on pressing issues, and even after the terrorist attack on 9/11 and calls for national unity, this remained treacherous territory. The terrorism battle raised new challenges. The president's appeal for congressional help in tightening domestic security had led to the introduction of the (United and Strengthening America) USA PATRIOT Act, which gave federal security agencies broad new powers.

Both sides of the aisle agreed that a tightening of security loopholes was crucial, and Daschle and Leahy joined Lott and Hatch to develop the Senate version of the Patriot Act on October 4. The administration, hoping to ensure the bill's quick passage, sent Attorney General Ashcroft, a former Republican senator from Missouri, to the Hill to lobby his longtime colleagues. But there were strong points of disagreement. Daschle was determined to include firm money-laundering provisions in the bill to address the financial networks that aided the 9/11 hijackers, and he balked at the administration's hurried timetable for passage. Liberal Democrats, led by Senator Patrick Leahy (D-Vt.), the former prosecutor who now chaired the influential Senate Judiciary Committee, argued that the proposed expansion of the FBI's powers of search and surveillance would infringe on constitutional liberties.

The debate had led to late-night standoffs in late September and early October. On October 3, Ashcroft and Lott had publicly accused Leahy of reneging on a carefully constructed agreement. Leahy suddenly found himself attacked by conservative radio hosts, who mobilized call-in campaigns to his office. The headline of one conservative publication, *Human Events,* branded the calm senator "Osama's Enabler in Congress."[76]

It was against this backdrop that the anthrax-laden letter addressed to Daschle at 509 Hart Senate Office Building was dropped into a New Jersey mailbox, and passed through the Hamilton postal center on October 9. There was nothing about it to arouse suspicion, although it was sealed with a layer of Scotch tape.

The Senate passed its version of the Patriot Act two days later, on October 11, 2001.

On the morning the anthrax letter contaminated his personal office, Daschle spoke by telephone to President Bush and gave him what few details were available. Bush broke the news to the White House press corps just after noon during a photo opportunity with Italian Prime Minister Silvio Berlusconi in the White House Rose Garden. The president described the Daschle letter's contents and noted that the white powder appeared to have been contained within a sealed envelope.

After weeks of guarded official comments, the Bush administration had begun to voice concern that bin Laden and his al-Qaeda network were behind the anthrax cases. A few days before, Vice President Richard Cheney had acknowledged in a televised interview with PBS's Jim Lehrer that the al-Qaeda network had made known its intentions to develop biological weapons.

"We know that [bin Laden] has over the years tried to acquire weapons of mass destruction, both biological and chemical weapons. We know that he's trained people in his camps in Afghanistan, for example; we have copies of the manuals that they've actually used to train people with respect to how to deploy and use these kinds of substances. So, you start to piece it altogether. Again, we have not completed the investigation and maybe it's coincidence, but I must say I'm a skeptic," Cheney said.[77]

Bush raised the same issue in his White House comments. Calling bin Laden an "evil man," he said investigators were "watching every piece of evidence. We're making sure that we connect any dots that we have to find out who's doing this. I wouldn't put it past him, but we don't have hard evidence yet.

"The key thing for the American people is to be cautious about letters that come from somebody you may not know, unmarked letters, letters that look suspicious. And give those letters and packages to local law authorities."[78]

At that point, however, authorities had only begun to explore the possible anthrax trail through the postal system. In Florida, Special Agent Pesquera said that the FBI and CDC had tested some postal workers in three locations, but only as a precautionary measure. Two New Jersey physicians had called the state Department of Health and Senior Services (DHSS) after hearing about the Trenton-postmarked letter that ended up at NBC in Manhattan. Both doctors had been treating postal workers with skin lesions who feared they might have come in contact with the Brokaw letter. One was a carrier, Teresa Heller, who worked out of the Ewing postal facility; the other was Patrick O'Donnell, who worked at a regional processing and distribution center in Hamilton. Heller's physician had been so mystified by her black lesion that he took photos of it, along with a skin biopsy in late September, but he hadn't put the pieces together until USPS announced that the letters had gone through New Jersey.

That same day, New Jersey State epidemiologist Eddy Bresnitz and acting Health Commissioner George DiFerdinando participated in a nationwide conference call put together by HHS Secretary Thompson's office. No one mentioned a possible threat to postal workers. "We didn't think postal workers were at risk at that point. We weren't ignoring it," Bresnitz said. "This was something that was off everyone's radar because it hadn't happened before."[79]

FBI agents went out to interview the two New Jersey patients and their physicians, and on October 14, they picked up Heller's biopsy specimen and delivered it to the health department.

The state medical examiner looked at Heller's tissue on October 16, and informed Bresnitz that the bacteria in the sample were consistent with anthrax. The lab packed the sample for shipment to Zaki's CDC laboratory for more specific testing.

At noon the next day, Bresnitz spoke to employees at the Hamilton plant. The workers had grown concerned that they might have been exposed to anthrax from the Daschle and New York letters. Bresnitz, citing guidance from experts at the CDC, instructed them that the chances of anthrax escaping a sealed envelope were "slim to none."

The workers asked for advice: Should they take antibiotics? The whole nation seemed to be rushing to drugstores to stockpile Cipro.

Bresnitz said there was no need for antibiotics. "We didn't have a case of anthrax among postal workers, so there was no recommendation to be made," he said later.

On October 18, the state got the results of CDC lab tests: Heller definitely had cutaneous anthrax, and the other postal worker was listed as a suspected case, the only diagnosis CDC could give without a skin biopsy.

Bresnitz received the news from Atlanta early that morning and drove into work knowing he was about to face the historic implications of announcing to the public New Jersey's first anthrax case. He and DiFerdinando had anxiously agreed that they needed to have a Plan B ready to roll out in case they found themselves facing a real anthrax attack. When Bresnitz pulled into the parking lot near his office in the Trenton state complex, he happened to see his colleague.

"Okay, George," he told him grimly. "Plan B."[80]

In a meeting that morning, Bresnitz and DiFerdinando pondered what to do about the worried Hamilton USPS employees. Heller's diagnosis made clear that enough spores could leak out of a sealed envelope to cause a cutaneous infection. "We had to make sure to protect the other workers," Bresnitz said.[81]

The two drove separately to the Hamilton facility to talk with postal officials and union representatives in the cafeteria. Their cellphones jangled for the entire ride as they called each other to talk strategy and run news questions by the CDC. When DiFerdinando asked if they should recommend that the workers take Cipro, the CDC doctors insisted that there was no reason to place the workforce on antibiotics for a case of cutaneous anthrax. The medication could guard against the inhaled form, but there was no evidence to suggest it would prevent a worker from contracting skin anthrax if he or she touched a contaminated letter.

Anxiety rippled through the group of workers. Bresnitz had just told them a few days before that they had nothing to worry about, and clearly he had been wrong. The postal unions pushed for proactive measures—even Hamilton Mayor Glen Gilmore wanted nasal swabbing done on the whole workforce. But in the frenzy surrounding the Florida and New York letters, the efficacy of the nasal swab had come into question. "A negative test doesn't mean they don't have anthrax; and a positive test doesn't mean they have anthrax, it just means they've been exposed," Bresnitz said. "You have to do it soon after exposure; otherwise people blow their noses and after two days there's nothing in the nose." Eventually, physicians at the Robert Wood Johnson University Hospital who were treating postal workers ordered swabs anyway.

The test proved useless for showing the extent of worker exposure. "Out of the thirteen hundred people who were tested," Bresnitz said, "not a single one of those people were positive." Authorities later estimated that eleven hundred people had been exposed.

Plant officials temporarily shut down the production line so that the FBI could take samples and health officials could survey the public areas. Bresnitz went to inspect the lines, not knowing that the facility was heavily contaminated and that he, too, would soon be on Cipro.

DiFerdinando rushed off to hold a press conference with acting New Jersey governor Donald DiFrancesco and other officials.

He explained that, based on CDC advice, the Hamilton workers would not be placed on antibiotics. A reporter asked a pointed question: Should someone who had passed through Hamilton's public areas be worried about exposure?

The health commissioner, thrown by the question, paused for what seemed like a long time. "I didn't feel like I could say zero risk," he said later. "So I said, 'If you ever feel like you have any skin problem, you should see your doctor.' I didn't deny that the public could be exposed. I carefully answered the question, and Eddy [Bresnitz] knew what I'd said: I'd basically said yes."[82]

Within days, both Bresnitz and DiFerdinando began to wonder if the CDC was passing along bad advice. Bresnitz recalled his own assurances to workers that anthrax could not escape from a sealed envelope. Could they have been wrong?

"That taught me a lesson," he said later. "Not everything the CDC says is necessarily accurate. Because they were flying by the seat of their pants, too."[83]

The FBI called Ezzell on October 15 to alert him that evidence would be brought from the Daschle crime scene straight to USAM-

RIID for testing. Ezzell's team knew that the stakes for quick and accurate analysis would be exceedingly high. Since the lab had been designed to process only several dozen samples a month, the lab staff had been stretched to the breaking point. It had grown from eight employees to eighty, borrowed from other USAMRIID divisions and sister agencies to handle up to seven hundred samples a day.

With the Daschle mailing, the murderer or murderers using anthrax to terrorize America had made another bold leap, from targeting a tight circle of media outlets in two states to provoking terror in the epicenter of the United States government. The letter ratcheted up the criminal case from the apparently unintentional victimization of an unlucky Florida photojournalist to attempted murder of one of the nation's most visible elected officials.

The FBI team pulled into Fort Detrick's gates bearing sealed containers and layers of Ziploc bags. The containers held the rapid assays performed by first responders at the Hart Building site. These tests had registered positive, but the CDC had repeatedly warned that their scientific accuracy had not been absolutely established.

Ezzell's team would have to account for the next level of analysis. First, they would use two more sophisticated tests that took about an hour each—known as rapid real-time molecular tests and fluorescent antibody assays. The most authoritative analysis, however, would require cultures that took his lab fourteen hours to process. Patience would yield the truth.

After the chain-of-custody paperwork was signed, giving USAMRIID legal possession of the letter and its contents, Ezzell and his team, following the usual protocol, began work in a BSL2 laboratory under a safety hood. Wearing a mask and gloves, Ezzell at this point had confidence in his own protection. He had been vaccinated so many times that he considered himself virtually anthrax-proof.

First, the team dealt with the sealed canisters, the assays wet with chemicals taken from the crime scene. They removed the contents and drained off some of the liquid to use for further testing.

Their concern heightened when they began examining the Trenton-postmarked envelope and its contents. The team had been trained to conduct risk analysis on suspicious packages and envelopes and to take extra precautions if they found anything even slightly alarming. Ezzell focused on the powdery material packed inside, shifting with movement. He told the team that for added protection they would step up laboratory safeguards before opening the letter and its inner layers of packaging.

The Ziploc bags were placed in a secure isolation chamber, known as an air lock, located outside the BSL3 laboratory. An FBI agent guarded the bags until Ezzell and his colleagues could get inside.

Using his identification badge, Ezzell entered the laboratory's outer change room, removed his street clothes, and pulled on green surgical scrubs. To access the lab, he had to pass through another security checkpoint by punching his personal ID number into a keypad. Once inside, he changed into lab shoes and made his way over to the biological safety cabinet. The FBI agent watched from outside a glass window.

Ezzell's first task was to remove the envelope's contents and photograph them with a digital camera. The FBI's anthrax squad planned to release to the public photos of the letter and its envelope, hoping that someone would recognize the peculiar script or the wording.

First, Ezzell thoroughly cleaned the cabinet, rinsing it with bleach and distilled water. He could see the fine powder dispersed inside the plastic bags and knew that his cabinet could become contaminated unless he took special precautions. He lined its bottom with a layer of bleach-soaked paper towels. That, he thought, would keep the spores under control.

Wearing multiple layers of latex gloves covered by sleeve protectors, Ezzell propped up the envelope against the back of the cabinet, forming a kind of artist's easel that would allow him to photograph the full image. He moved back to focus the camera. That was when he noticed it: the bleach had wicked up through the dry panels. The bottom of the envelope, an important piece of criminal evidence, had become smudged with bleach solution.

Oh my God, what have I done? Ezzell thought. Worried that he might have tainted the evidence but unable to undo the damage, he carried on. He began slowly removing the letter from its envelope, relieved to find it undamaged. But as he worked, he noticed a bit of white powder tucked into one of the letter's folds. Almost as soon as he saw it, the powder dispersed, spreading invisibly through the safety cabinet.

After years of researching anthrax, he had never seen the bacteria in its weaponized form—the type made in Fort Detrick's Building 470 during the 1940s and '50s—a material that could blanket a city or annihilate an enemy. This was a powder so virulent that normal laboratory rules did not apply. Both he and his team could be at risk despite their precautions.

"I already knew that it was anthrax spore powder, and then I saw the form of it, and I said, 'After all these years of looking, here it is. This is the real thing, in the right form,'" he recalled.

Ezzell kept his fear suppressed, determined to finish the job carefully. He finished pulling the letter out, then coaxed the spore powder back into its Ziploc bag. The letter went into a sterile baggie, and he sealed the envelope in another one. Then he wrapped both in plastic bags that had been decontaminated with bleach solution.

Ezzell took photographs of the letter and envelope, then took the sterile bags to the lab's glass window and held them up so that FBI agent Darin Steele, waiting outside, could snap another set of photographs. Those first images of the Daschle letter, made through a panel of glass, with Ezzell's gloved hand visible from behind, were seen around the globe.

"I had worked a long day . . . and sometimes I would end up pretty haggard. They told me later, 'We would have shown more of you in the picture, but you were looking pretty tired, so we decided to show just the letter,'" he said, laughing.

When his work was completed, Ezzell deconned and left the laboratory. He was too unpretentious to think of his actions as heroic; usually, he stuffed army commendations and awards into a desk drawer. Yet if the FBI hoped to catch its killer, the risks taken by Ezzell and his team would be critical to building a criminal case.

To protect himself, Ezzell started antibiotics to guard against infection. He also took another precaution. Ezzell went to a sink and mixed a solution of diluted bleach. Bracing himself, he lifted it to his nose and took a deep snort. The pain that surged through his sinuses almost knocked him to the ground, but he could not stand the thought of carrying anthrax spores in his nostrils.

Ezzell's boss, Col. Erik Henchal, the chief of USAMRIID's Diagnostics Systems Division, waited outside, and Ezzell gave him the distressing analysis.

"In all of my training and talking to people like Bill Patrick, looking at what's required for aerosolized, easily dispersed powder, this stuff came close to the criteria. This is something that was prepared."

Later, in one of the regular interagency conference calls, Ezzell described what he had seen when he looked into the Daschle letter. He used the term *weaponized anthrax*.

That night, a friend who worked for the CIA woke him from a deep sleep to tell him that his assessment of "weaponized" anthrax in the Daschle letter had been passed on to the president of the United States.

John Ezzell's comment ignited a fire that spread around the world. The term *weaponized,* which was quickly leaked to the media, implied to many that the material had been prepared as an agent of war by a hostile nation. This fanned fears that bin Laden's followers were behind the attacks, possibly using anthrax developed in Iraq.

Ezzell had not meant to imply that a hostile nation was behind the U.S. letters. He had simply been describing the astoundingly fine aerosol material—more easily dispersed than any he had ever seen. The anthrax in the New York letters had been the consistency of wet dog chow, but the powdery form of the Daschle letter's anthrax suggested a perfection developed through laboratory practice. Ezzell had no idea if international terrorists had created what he saw in his safety cabinet, but he knew one thing immediately: a professional had made the anthrax, most likely one who had access to a sophisticated lab.

Major General John Parker, then the commanding general of USAMRIID, found himself having to explain Ezzell's weaponization comment in hearings on Capitol Hill. Parker explained that his scientists had "revisited the term 'weaponized' and decided the terms 'professionally done' and 'energetic'" were more appropriate descriptions, "in lieu of any real familiarity with weaponized anthrax."[84]

Ezzell was stunned by the furor he had created, but felt no regrets.

In his view, he needed to sound the alarm that the powder contained in the Daschle letter was a killer, capable of leaving a trail of bodies in its path.

There also was growing concern about the strain. Tests of the anthrax samples in Florida and New York had found them virtually indistinguishable, and the preliminary DNA analysis of all of the anthrax had identified it as the Ames strain. It was becoming increasingly likely that whoever committed these crimes did so with material linked in some way to a U.S. military lab.

At Iowa State University, panicked researchers dealt with the speculation by incinerating the school's anthrax isolates. Later, questions arose about how the FBI could have allowed the destruction of material that could have proven critical in tracing the anthrax used in the attacks.

The FBI also came under attack. In New York, complaints surfaced about its lack of diligence in the NBC case. When network officials first reported the suspicious September 25 letter to the overwhelmed New York FBI field office, the response was halfhearted. The letter and its contents were not tested until O'Connor's doctor, Richard Fried, took a skin biopsy and referred her to a specialist. On October 6, two weeks later, someone finally called the New York City Health Department, and officials there were upset that the FBI had done nothing with the letter. New York supervising agent Barry W. Mawn conceded that there had been lapses. Concerned that his FBI office in Lower Manhattan could be a target if terrorist hijackers struck again, Mawn had moved his operation into makeshift quarters in a concrete FBI garage. Virtually everyone there was consumed with the investigation of 9/11, leaving few hands free to look into what seemed like unrelated health issues.

Around the country, similar concerns surfaced. Martin Hugh-Jones, whose Louisiana State laboratories housed one of the world's most extensive collections of anthrax isolates, waited for several weeks before agents came to inspect his records of anthrax transfers and sample types. Near Fort Detrick, retired veterans of the offensive biological weapons program griped that the FBI had not seen fit to contact them early on for technical assistance and advice. Even Bill Patrick heard nothing.

Octogenarian James R. E. Smith, who had worked in Detrick's offensive weapons laboratories, got so upset that he hadn't been contacted by the FBI that he wrote a three-page letter to Homeland Security Chief Tom Ridge describing a prime suspect.[85] "I described this individual, his home, his education, and in the very last paragraph, I said, 'This individual is me,'" Smith said. The FBI finally paid a visit.

The FBI also neglected to contact some victims of the attacks, like Huden of the *New York Post,* who said she was not interviewed about suspicious mail.

FBI officials tried to answer some of these criticisms and project an aggressive image on October 16, a day after the Daschle letter surfaced. FBI Director Robert S. Mueller and Attorney General Ashcroft held a press conference to announce the arrest of a Connecticut man for his role in an anthrax hoax at that state's Department of Environmental Resources. The incident caused the evacuation of eight hundred people, including twelve who had to be washed down with bleach solution.

Ashcroft and Mueller defended the FBI's performance in New York, and declined to address questions about the similarities of the anthrax used in the scattered attacks.

Ashcroft had been the engine behind a government-wide tightening of the release of information to the public, justified, he said, by the unprecedented threats to national security. A reporter asked him if more openness was called for in the anthrax case because of its public health implications. Ashcroft said that while sharing reliable information was important for prevention purposes, the criminal case was another matter. [86]

Ashcroft's cautious words were well chosen. The FBI was still recovering from an embarrassing string of missteps that had called into question not just its investigative rigor but its competence.

In February 2001, then FBI Director Louis Freeh had faced one of the bureau's most humiliating moments, the arrest of veteran agent Robert Philip Hanssen on espionage charges. Revelations of a spy working undetected for years within the nation's chief law enforcement agency had touched off a furor, resurrecting discussion of FBI mistakes in the investigation of Oklahoma City bomber Timothy

McVeigh and false allegations made against nuclear scientist Wen Ho Lee and Richard Jewell, a suspect in the 1996 Atlanta Summer Olympics bombings.

Freeh announced in May that he would leave the FBI for the private sector, and Bush chose Mueller, a former U.S. attorney in San Francisco, to take on the Herculean task of rebuilding confidence in the agency. He was confirmed just weeks before 9/11.

Although Mueller often stood in the spotlight during the anthrax case, behind the scenes, others were doing crucial work: hundreds of agents had been detailed to the case, many of them already working in distant lands on the 9/11 hijacking case when the anthrax attacks began. This proved fortuitous. Some of the most intriguing hoax letters had originated in other countries and required careful tracking.

In keeping with FBI tradition, management of the day-to-day anthrax investigation was headed in the field offices where the attacks occurred. This approach worked well when a case was geographically limited, but coordination and communication became challenging when a criminal matter involved more than one FBI jurisdiction. The offices in Florida and New York had investigations well under way by the time the Daschle letter surfaced in Washington. Van A. Harp, who was in charge of the important Washington, D.C., field office, assumed responsibility for investigating the case on Capitol Hill, and ultimately was named to coordinate the national anthrax investigation, dubbed Amerithrax.

Harp was an old-school FBI agent who had made his name prosecuting the Mafia in Cleveland, Ohio. Impressively fit, he had a dimpled, welcoming smile that could evaporate into a cold stare, and his piercing gaze seemed to function as a built-in lie detector. Located near Washington's Judiciary Square, Harp's field office handled some of the FBI's highest-profile cases, including criminal matters that arose from Capitol Hill and Washington's federal bureaucracy.

In its investigative strategy for the anthrax matter, the FBI applied a lesson from the Unabomber case, in which it spent thirteen years searching for a mysterious letter bomber only to eventually learn killer Ted Kaczynski's identity from his brother, David. The break in the investigation developed after the FBI allowed media outlets to print a long screed Ted Kaczynski had sent to the *Washington Post* and the *New York Times*. Reading the coverage, David Kaczynski immediately recognized the bizarre thought process and peculiar phrasing from letters he had received over the years from his reclusive brother.

In the anthrax case, investigators hoped that they could trigger public response by releasing pictures made during the USAMRIID letter opening and some details of the letters. "In the past, the public has helped the FBI solve high-profile investigations that involved writings by coming forward to identify the author, either by how he wrote or by what he wrote. We are asking for the public's help here again in the same way," the bureau said in a November 9 press release.[87]

The same release contained a linguistic and behavioral analysis of the letters, which concluded that the sender was most likely an adult white male with some scientific background who was likely to be in little contact with the public. The killer, most likely a "nonconfrontational person" appeared to have some familiarity with Trenton, New Jersey, and targeted his letter against Daschle by looking up the correct office address and a specific eight-digit zip code assigned to the senator.

The FBI urged the public to be alert for individuals fitting the description who "may have become more secretive and exhibited an unusual pattern of activity" after 9/11. "He also may have started taking antibiotics unexpectedly," it said.

In truth, the help the FBI desperately needed was that of the scientific community. It needed to solicit tips about suspicious characters lurking around government and university laboratories, about missing supplies and disturbed scientists with grievances against the U.S. government. The mailings had offered few clues, which the bureau outlined:

EVIDENCE DESCRIPTION
Letter 1
One page, hand-printed letter
Transmittal envelope, also similarly hand printed
Addressed to "NBC TV—Tom Brokaw"—No return address
Postmarked Trenton, NJ 09/18/2001 (Tues.)

Letter 2
One page, hand-printed letter
Transmittal envelope, also similarly hand printed
Addressed to "NY Post"—No return address
Postmarked Trenton, NJ 09/18/2001 (Tues.)
Letter 3
One page, hand-printed letter
Transmittal envelope, also similarly hand printed
Addressed to "Senator Daschle—509 Hart Senate Office Building"
Return address—"4th Grade, Greendale School, Franklin Park, NJ"
Return address zip code—"08852"
Postmarked Trenton, NJ 10/09/2001 (Tues.)

Since opening the Daschle envelope, John Ezzell would bolt awake in the middle of the night, worrying about what would happen if anthrax powder made its way into American households. This stuff was coming through the mail, he told himself. Everyone was vulnerable. What could the average person do to protect himself?

Every day, there were new incidents around the globe. Suspicious letters showed up in France, Italy, Africa, at the stock exchange in London. Another letter appeared in New York at the Manhattan office of Governor George Pataki, although there was not enough material left in the envelope to determine if it was anthrax. All of the scares had been tied to the mail.

Ezzell's mind raced, the scientist in him trying to find a solution. He knew that heat could destroy anthrax spores. In fact, it was well documented in the scientific literature that two hours of exposure to dry heat at 320 degrees Fahrenheit would kill spores. Was there a simple method the average person could use before opening mail to kill any anthrax without destroying the contents?

Ezzell, an accomplished cook, decided to try his kitchen oven. Usually, he would stand over the stove stirring homemade soups or testing new variations of his prizewinning pickle recipes. This time, he set the temperature to 320 degrees and waited for the oven to heat. He inserted a package of daily mail—sealed envelopes of various sizes and types, plastic-encased magazines, newsprint circulars. He waited two hours and opened the oven.

Voilà! The stamps remained firmly in place; the envelopes were still solidly sealed, though the plastic windows of some showed slight shrinkage. Glossy magazines had a slightly burned appearance but were readable. The process let off some smelly fumes, the scent of burning paper, but Ezzell fixed this by turning on the exhaust fan.

Over days and restless nights, he perfected the technique. The process worked much better if the batch of mail was placed inside a turkey-basting bag or foil container. A careful person could take the bag to the mailbox, dump the contents inside, secure with a twist-tie and easily pop it into the oven.

He tried the process in a toaster oven, but it was too risky; the temperature could not be accurately controlled. A microwave oven tended to ignite certain types of paper. Fire extinguishers should be readily available, he decided.

Finally, after testing the oven technique in a BSL3 lab on some spores from the Daschle letter, Ezzell sat down to write about his findings in a paper called "Procedure for Killing *Bacillus Anthracis* Spores in Mail." He laid out the detailed instructions, including a disclaimer: "The author assumes no responsibility for loss of plastic items (including credit cards), fires, odors or other damage," he wrote. He explained how to sterilize plastic wrappers with diluted bleach and recommended wearing latex gloves.

"While there may be opportunities for fine-tuning the process, the advantages of this approach are, that the process is low-tech, immediately available, and can be performed in residences or offices. It is based on firm scientific data with respect to temperature and time required for killing *Bacillus anthracis* spores and with respect to initial experiments which have shown that spores from Senator Daschle's office are killed well within the two-hour heating period," he wrote.[88]

Ezzell began distributing the guide to friends and fellow worshipers at his Methodist church. Someone posted it on the Internet.

It seemed funny that after a lifetime of scientific endeavor, studying the fine points of an obscure and mystifying bacteria, this would be his most practical contribution to the common good. Like a chef in a Betty Crocker cook-off, he had created the homemaker's guide to baking anthrax sealed in the U.S. mail.

CHAPTER 10

Neither Snow, Nor Rain

October 18, 2001
WASHINGTON, D.C.

Three days after news of the Daschle letter spread fear across Capitol Hill, Postmaster General John Potter embarked on a damage control mission. His first stop was the White House, where Tom Ridge was assembling a collection of federal luminaries for a press conference intended to deliver a coordinated message of reassurance to the American public. After that, Potter headed to the U.S. Postal Service's Brentwood distribution center to try to ease the mounting concerns of the nation's 800,000 postal workers and to announce a $1 million reward for information leading to the capture of those responsible for the anthrax mailings. The reward would climb to $2.5 million over the coming months.

Focusing attention on Brentwood's still-churning operations, Potter and high-ranking FBI officials stood before television cameras emphasizing their determination to keep the mail flowing and to catch the anthrax culprit. The fear surrounding the Daschle letter had gained worldwide attention and threatened to get worse. Already, there were suggestions that the USPS should curtail operations in certain cities until the causes and scope of the anthrax infections were better understood. Estimates surfaced saying it would cost the USPS billions of dollars to close down major distribution centers. The idea was heretical to Potter and the USPS Board of Governors, who considered the mail the lifeblood of the nation.

But just that morning, the CDC had called New Jersey officials to confirm Teresa Heller's skin anthrax case and then passed along word through the federal chain of command to Potter's office. If worker health issues escalated, or if the frightened public abandoned the USPS for alternative mail carriers, it would only worsen a looming financial crisis. Long criticized on Capitol Hill for inefficient management, the USPS had been hit hard by the flagging national economy, losing a staggering $1.7 billion that year. Potter, who had taken over in June as the nation's postmaster general, had been handed the unenviable task of bailing out the sinking ship. He was determined to cut costs by $900 million and improve efficiency in the delivery of more than 200 billion pieces of mail each year.

A self-described "people person," Potter had spent twenty-four years with the postal service, starting as a postal clerk and climbing his way to a king-size office suite at the USPS headquarters on Washington's L'Enfant Plaza promenade.[89] With a hefty build and a shock of white hair contrasting sharply with his thick, dark eyebrows, he looked out of his element fretting over paperwork. It was easier to imagine him hanging out with the guys at Brentwood's cafeteria, laughing and talking trash, than managing one of the most challenging agencies in the federal bureaucracy, with long-standing labor issues and an operating budget of $54 billion.

Potter had enough experience wrestling with postal unions to know that something needed to be done quickly to soothe postal workers nationwide, especially in the anthrax-affected areas of New York, New Jersey, Florida, and Washington. Almost every day, work in a postal facility somewhere in America was halted because of a suspicious box or letter or a sudden burst of dust or powder. Tests in a Boca Raton, Florida, post office that serviced American Media Inc. had turned up minuscule quantities of spores, and employees were placed on antibiotics, although the building remained open for business.

In New Jersey, Heller's skin lesion was the first visible sign that anthrax was threatening postal employees as they went about their daily business. New Jersey State officials had huddled that morning to consider the awful possibility that spores had seeped not just onto a carrier's hand but also into neighborhood post offices and the huge Hamilton processing center. They prepared to meet with union representatives who were beginning to imagine the same scenario.

Postal employees, from rural carriers in the heartland to assembly-line workers in huge facilities like Brentwood, had begun demanding that management protect them from anthrax. From the outset, there were racial implications. The postal service's rank-and-file workers were predominantly African-American, and the most outspoken among them suggested that they were being subjected to a double standard. Reports of prompt, attentive medical care for the largely white workforce of the U.S. Senate, and the extensive environmental testing of Capitol Hill, contrasted poorly with the lackadaisical response offered to the postal workers.

When the Daschle letter surfaced, postal officials immediately knew that it would have passed through the processing center in Hamilton and then through Brentwood, the central distribution point for all U.S. government mail in the Washington area. From there, it had traveled to a postal station on P Street NW, then to the Senate mail distribution center in the Dirksen Senate Office Building, and finally through a hallway to the adjacent Hart Building. Along the way, the envelope had rested in various sorting trays, trucks, and mailbags, any of which could have been contaminated by wayward anthrax spores.

But Potter felt that his employees were in no real danger. His executive staff had been consulting with doctors from the CDC and the D.C. Department of Public Health since the Daschle letter surfaced, and the CDC's top infectious disease specialists had concluded that there was no reason to start 2,400 Brentwood employees on preventive antibiotics.

Potter didn't question this advice, and after discussion with Koplan, he didn't seek further medical opinions.

Azeezaly S. Jaffer, the USPS vice president for public affairs and communications, suggested holding the televised press conference inside the Brentwood facility; if workers were frightened, it would ease their minds to see the top postal brass and other federal and local officials standing inside the building.

But the calculated symbolism of the Brentwood event turned out to be broader than its organizers imagined. The press conference would stand as an illustration of the government's bumbling response to the dangers posed by the anthrax-laden letters. Within days of the conference, the genial, handshaking Potter and others who had unwittingly walked into the ground zero of the anthrax affair would find themselves taking Cipro to ward off infection. With a taped envelope and a thirty-four-cent stamp, the letter's sender had outsmarted the government's top scientists, its doctors, and the seasoned bureaucrats running a critical part of the nation's infrastructure.

Earlier that morning, Potter had stopped by the White House press briefing featuring Tom Ridge's debut as Bush's director of homeland security. Ridge's status had been formalized on October 8, with orders from the president to "provide the American people as much accurate information as we can, as soon as we can," Ridge said. Taking the advice to heart, he had gathered together almost every agency official involved in the anthrax case, telling reporters, "Our government is more coordinated than ever."[90]

Ashcroft and Mueller appeared to discuss the criminal investigation. The CDC's Mitch Cohen, who was still detailed to FBI headquarters in Washington, joined an ever-expanding fold of public health officials, Surgeon General David Satcher, and Deputy Surgeon General Kenneth Moritsugu. With Koplan working in Atlanta, Cohen had been notified that morning to show up at the White House to represent his agency.

Ridge had just been briefed about Heller's diagnosis, and he mentioned to reporters that authorities might soon give notice of a sixth infection. Seeking to put the events in perspective, he noted, "thousands and thousands of people have been treated for anthrax exposure, and thousands of environmental samples have been taken as well. Yet only five people have tested positive at this time for anthrax."

On Capitol Hill, he said, more than three thousand nasal swabs had been taken, with thirty-one preliminary positives, and all those tested had been offered precautionary antibiotics. But Ridge made no reference to postal workers' fears, or to concerns about possible exposure along the path of the Daschle letter.

Potter stepped forward to reaffirm the safety of the U.S. mail. He said a postcard would be sent to each American home, explaining the warning signs of suspicious mailings.

"My message to the American public is to remain calm, be vigilant, be aware of what you get in the mail," Potter said. "And as the president has said, heightened awareness right now is what we are asking all of America to do. Yes, we believe that the mail is safe. It is very safe if you are prudent, and if you follow the simple directions that we're asking you to follow."[91]

Satcher detailed the preventive role of antibiotics in treating anthrax, stressing repeatedly that the government had ample supplies to treat anyone who feared he or she had been exposed. "We are delivering the appropriate medications to those who need it, and we are erring on the side of caution in making health care available to those who may have been exposed to anthrax spores," he reported.[92]

The White House press corps listened respectfully as the government officials congratulated themselves for jobs well done and told the nation that the communication gaps and slips in coordination were a thing of the past.

When the floor opened for press questions, the tone was reserved. No one asked about the possible exposure of the postal workers who handled the Daschle letter or why the free-flow of antibiotics on Capitol Hill stopped abruptly at Brentwood's doors.

The press conference at Brentwood attracted reporters from around the world, but postal workers went about their business, pausing, as Leroy Richmond did briefly that morning, to check out the television lights and cameras. Potter and his aides had succeeded: Brentwood employees seemed comforted to see the cluster of government bigwigs standing in the eye of a gathering storm.

FBI Deputy Director Thomas Pickard and Chief Postal Inspector Kenneth Weaver, who told reporters that he would "ensure the mail continues to be a welcome visitor in every home and business in the nation," announced the reward. "We know there is someone, somewhere who saw something or who knows something about these letters."[93]

But behind the confident words were signs of mounting frustration among federal law enforcement personnel. Weaver and Pickard used the media opportunity to promote the effective partnership between federal law enforcement and the popular television show *America's Most Wanted,* whose host, John Walsh, stood with them.

Postal officials had found Walsh's show effective in the hunt for missing children, and it seemed only natural to turn to the detective's program to help solve the anthrax mystery. Walsh noted that in these troubling times the collaboration, which had helped nab 683 fugitives, was "more important than ever."[94]

When the news conference was over, Potter stayed behind to walk through the plant and glad-hand the workers. He stayed longer than expected, strolling by the sorting machines, including the soon-to-be-infamous Machine No. 17, and shaking hands with workers in the heavily contaminated government mail area.

Potter would later cite the press conference as proof of the USPS's complete confidence in the advice of CDC health officials and of the postal service's innocence in a series of regrettable decisions. In choosing the Brentwood building as the setting for the event, the USPS had relied on the firm assurances of CDC doctors that the plant was safe. Later, with the advantage of hindsight, Potter, a husband and father, said that if he had known Brentwood was contaminated, there was no chance that he would have placed himself in harm's way.

That weekend, as Leroy Richmond was declining in a Fairfax, Virginia, hospital, Potter drove over to D.C. General Hospital in southeast Washington to join hundreds of other postal workers who had been waiting for hours to receive nasal swabs and free antibiotics. He was just like the others, he later joked, except that when he saw the long lines, he pulled rank and cut through to the front.

Since the first anthrax attack, postal officials had begun consulting with the CDC on how to educate workers about the germ and its medical effects. On October 17, Potter and his aides asked a more pointed medical question: Should postal workers along the Daschle letter's trail be placed on preventive antibiotics?

Even in such busy times, CDC doctors tried to approach each new question in a methodical, scientifically grounded manner. Antibiotics were not supposed to be doled out to patients without even a suggestion of anthrax exposure. Cipro was a powerful drug that would lose its effectiveness with overuse, and some within the CDC had become convinced of the need to prescribe more cautiously, based on individual case analysis.

There was no evidence yet that a single spore had leaked into Brentwood, and the CDC team had not even begun to do sampling there. It was still working its way through the overwhelming demands of Capitol Hill.

"It was like going into the water and getting slammed by a wave," recalled Koplan. "You get hit quickly, you get wet, and you keep moving on. The call would come: 'We got a letter here, we got a case here.' And it would be like, *wham!* That shock of the first cold and wetness, and then you realize you're going in the water, you just keep moving and dive right in. During the course of the event, you don't stop and say, 'Wow! What are the implications of this on a grand scale?' You do stop and say, 'How is this different? Is there new information? What light does this shed on what we've just seen in three, two, one other places?' Indeed, that was the standard feature of this. In each one of these events, you could look back on what had gone on the previous days and maybe gain some better insight. Maybe not."[95]

The Daschle letter made it obvious that the sender of the anthrax letters was trying to make a political statement. "It was no longer just a media focus," Koplan said. The anthrax threat had the ability to terrorize "an important element of the country's leadership."

Dr. Rima Khabbaz, the deputy director of the Division of Viral and Rickettsial Diseases in the National Center for Infectious Diseases, had been sent to Washington to head the CDC investigation. Previously, she had traveled around the world investigating outbreaks of exotic illnesses for the CDC, including a 1995 Ebola virus case in Zaire. Still, she was not prepared for the frenetic atmosphere of Washington in crisis.

Khabbaz knew that she was headed into a maelstrom of multiple agencies with competing jurisdictions, demands, and egos. She had to communicate not only with her superiors at the CDC but with leaders at HHS, the D.C. Health Department and its local bureaucracy, the National Institutes of Health, the surgeon general's office, the Senate's attending physician, and ultimately the state health departments in Maryland and Virginia.

The letter had come to the office of one of the Senate's most important figures, who wanted to be informed of every development and was the focus of constant media attention. Daschle praised the extensive medical team treating Hill staffers, saying they faced a challenge "unprecedented in American history. To a person, they have responded admirably. Their poise, professionalism, and compassion have been a comfort to us all."[96]

Not to be outdone, Senate Minority Leader Trent Lott, of Mississippi, asked Frist, the Tennessee doctor, to brief the Republican side. Frist, who was speaking at a bioterrorism conference in Tennessee when news of the Daschle letter broke, rushed back to Washington and played a major role in medical consultations, using his office Web site as a forum for medical information about anthrax. He, too, appeared on the major television networks, assuring Americans that "the system is working. People are working together in a harmonious, almost symphonic way."[97]

Koplan observed the spectacle from his office in Atlanta. "Whenever you looked at a press briefing, or a conference or an announcement during that period, the stage was just filled. I certainly found it ironic that Rima, who is probably the best person suited to do the investigation—that's why she was there—was often eight people over in the third back row. But then a question would come up about something substantive related to the event and you'd see them dragging her, pushing her forward, to get in front of the mike. It was an interesting choreography."[98]

Khabbaz's team faced new difficulties at every pass. The first tally of anthrax exposures, thirty-two, only hinted at the size of the problem, and the magnitude of the fear. Long lines of Hill staffers waited for testing, and virtually everyone who had set foot in the contaminated building was demanding antibiotics. In one day, one thousand people lined up to be tested.

On October 18, the CDC issued an alert recommending that anyone who was on the fifth or sixth floor of the Hart Building on October 15, between 9 A.M. and 7 P.M., receive sixty days of antibiotics to guard against suspected exposure.

Cautious leaders in the House of Representatives recessed the chamber and closed office operations. But the Daschle-led Senate, in a statement to would-be terrorists, decided that the chamber must remain open, despite the inconvenience of having its three office buildings closed for testing.

As feared, tests showed that the Senate mailroom contained anthrax spores. Next, Khabbaz's team would follow the Daschle letter's trail back to the P Street post office, then to Brentwood.

When the question came from USPS officials about the need for antibiotics, Khabbaz was a day or two away from extensive testing at Brentwood. The workers' urgency and the developments in Trenton had forced the issue of whether Cipro should be given out before rigorous testing was conducted. When the Daschle envelope passed through Brentwood, still tightly sealed, what were the chances it had leaked enough spores to threaten a huge workforce and justify the wide use of Cipro?

Though they did not know it at the time, one set of answers that might have guided these discussions sat in an unattended e-mail inbox at the CDC headquarters.

Although there had been numerous anthrax hoaxes sent through the mail in the late 1990s, no researcher had analyzed how a real anthrax-laced letter might spread spores through a crowded office building, much less through the postal system.

In February 1999, the CDC's *Morbidity and Mortality Weekly Report* summarized the investigations into hoax letters in Indiana, Kentucky, Tennessee, and California, and their possible public health implications. The report was studded with such mild conclusions as: "The public health response to bioterrorism requires communication and coordination with first responders and law enforcement officials."[99]

It did not suggest drugs as a first response to potential exposure. Instead, it urged public health officials to give those potentially exposed "information about the signs and symptoms of illnesses associated with the biological agent and about whom to contact and where to go should they develop illness."

The CDC doctors were unaware of a relevant study that had recently been conducted by military researchers in Canada, though a report describing it was sent by e-mail to a CDC official. The study—by scientists at DRDC Suffield, a Canadian government facility operated by Defense Research and Development Canada—showed that anthrax contamination could spread easily and thoroughly with the opening of a single letter. Completed in September 2001, it was initially classified, but a researcher e-mailed it to an official with the CDC's Laboratory Response Network around October 4, when the first anthrax letter surfaced. The report sat among the official's unread e-mails for at least three weeks.

The study concluded, "a lethal dose could be inhaled within seconds of opening an anthrax-spore-filled envelope." In the ten minutes it takes to respond to an emergency call, the study said, a victim could have inhaled many times the lethal dose. "In addition, the aerosol would quickly spread throughout the room so that other workers, depending on their exact locations and the directional air flow within the office, would likely inhale lethal doses. Envelopes with the open corners not specifically sealed could also pose a threat to individuals in the mail handling system."[100]

Bioweapons pioneer Bill Patrick had also reported on the potential effects of two grams of anthrax sent in the mail in a February 1999 study he did for Science Application International Corporation at the behest of Dr. Steven Hatfill, then employed by the company. Patrick said it was his understanding that the study had been sent to the CDC for use in formulating a pamphlet to teach first responders how to react to biological agents. It is unclear who at the CDC received the report, but it, too, failed to appear in the discussions about Brentwood.

It never dawned on the CDC officials that refined anthrax could simply seep through the pores of a standard envelope. This failure in reasoning would soon cast the CDC experts as "stupid," Koplan remembered, although months later, he would still find the concept of porous paper hard to grasp.[101]

In retrospect, Koplan said, it would be easy to say that public health authorities should have been more familiar with the postal process and should have spent more time analyzing the risk of a sealed envelope. "All I can say is, with what we know now, yes, we would view mail, the way it gets handled, and even sealed envelopes in a different framework," he said. "But we didn't know it then."

In Washington, Khabbaz passed along word to postal officials that her colleagues saw no need to medicate the complaining Brentwood workers. The USPS officials leaned on the CDC's opinion. As Potter recalled, "We were assured throughout this process . . . that our employees were not exposed."[102]

Based on this advice, Potter's staff proceeded with plans for the Brentwood press conference on October 18 and hired a contractor to take samples at Brentwood before Khabbaz's team arrived. The samplers took swabs from around the floor and near the sorting machines, retracing the likely path of the Daschle letter.

To the relief of Potter's staff and the CDC advisers, the contractor's preliminary tests were negative for anthrax. There was no need to close Brentwood, they reasoned, and no reason to fear widespread contamination of workers.

Their confidence was false, based on porous research methods. It would take several days and the outbreak of Leroy Richmond's illness to clarify the true risks.

CDC doctors had offered the same advice to New Jersey health officials. But when Teresa Heller was diagnosed with cutaneous anthrax, DiFerdinando and Bresnitz began to regret telling the Hamilton postal workers that they were safe. They had temporarily shut down the plant on October 18, but New Jersey–based USPS officials wanted to reopen the facility in time for the next morning's shift. To wrestle the question, they set up an 8:30 P.M. meeting at a Princeton hotel.

The CDC sent a small team from Philadelphia to offer technical assistance. A similar number of USPS officials tuned in by telephone or in person. "They weren't badgering us," DiFerdinando recalled, "but they were under this desire to do the job. Their job is to deliver mail."[103]

The dispute raged for hours. Around 1 A.M., DiFerdinando bolted awake; he had been dozing in his chair at the head of the table. "I just fell asleep; and I woke back up and they were still debating," he said.

DiFerdinando awoke from his nap with the realization that the situation called for decisive action. He looked around the table, listening to yet another argument from the postal officials about why the Hamilton plant should remain open.

"Hold it, hold it, hold it. I've figured this out. I don't know what you're going to do tomorrow," he said, pointing to the USPS officials. "But I know that if you're going to reopen the facility, we're not going to be standing there next to you."

Postal representatives stressed that they needed the backing of DiFerdinando's department to reopen. He was reminded of the widely covered press conference that very morning at Brentwood. DiFerdinando recalled being asked, "How can [you] stand there and say Hamilton is not safe when Brentwood is?"

But he held firm. "Once you close a building in an occupational exposure setting, you have to know it's safe before you reopen it. Here, we had closed the facility for convenience, to get the testing done. Once it was closed, I knew the answer. We couldn't open it back up until we knew it was safe."

The officials tumbled into bed around 3 A.M. on October 19, with the Hamilton plant closed for the foreseeable future. The next morning, they felt vindicated. Word reached their offices that a Hamilton worker from Bucks County, Pennsylvania, had been hospitalized with a nasty lesion. That brought to three the number of Hamilton workers who had contracted cutaneous anthrax.

Bresnitz and DiFerdinando revisited the question of administering antibiotics to the Hamilton workforce. The CDC continued to stress that there was no evidence that anthrax had become aerosolized within the Hamilton plant, and no tests had been conducted to narrow the scope of possible worker exposure. The only option was giving drugs to everyone, regardless of whether they needed them.

CDC officials, DiFerdinando recalled, did not want the New

Jersey Health Department to "muddy the waters by treating everyone. We got that advice from CDC and we decided, with all due respect, we were going to treat these people anyway. From a scientific standpoint, there was no absolute proof of aerosolization, but we went ahead because we didn't know what was going on."

On Friday, October 19, the New Jersey health authorities recommended that all workers at the Hamilton site receive an initial round of prophylactic antibiotics. Neither the CDC nor the USPS was represented at the news conference where the plan was announced. At the urging of the Hamilton mayor, postal workers went to the Robert Wood Johnson University Hospital for medication and testing. Postal officials later balked at paying the bill for their care.

"We didn't understand why Brentwood was still open and why people weren't being treated, but those were our opinions," DiFerdinando explained. "Others seemed to not believe that things could be aerosolized. And once we'd looked at the process, there were letters basically going through high-speed pasta machines. Why wouldn't powder squeeze out the back end? It's an envelope made of paper."

Over the next few days, with the Hamilton plant out of commission, the New Jersey officials still had to tend to Hamilton worker Norma Wallace, who had been infected by aerosolized anthrax before the production lines ground to a halt. Her illness prompted Bresnitz, DiFerdinando, and CDC officials to extend the recommendations for antibiotic treatment to sixty days.

As a series of horrific events unfolded in Washington, DiFerdinando and Bresnitz could not help but feel that their decision had most likely averted a broader tragedy. Wallace, bolstered by antibiotics and prompt diagnosis, eventually recovered from inhalation anthrax. Two workers at Brentwood were not so lucky.

CHAPTER 11

A Warning Ignored

October 20, 2001
FALLS CHURCH, VIRGINIA

Around 2 A.M., when she finally left her husband's bedside at Inova Fairfax Hospital, Susan Richmond frantically dialed the number for Rich's workstation at Brentwood, trying to remember anyone who might have worked near him.

"If he's this sick, and he works with other people, they could be sick, too," she told herself. She wanted to get through, to tell everyone to "get out of there and tell the supervisor."

She had begun telephoning Brentwood hours earlier after a blood test convinced doctors that Rich had contracted inhalation anthrax. Her first call was to the office of Brentwood manager Timothy Haney. The phone rang and rang until her call was switched to voice mail. Susan tried to contain her frustration as she left a panicked message. Her husband had just been diagnosed with anthrax. "You need to protect these people." Shut Brentwood down.

Angrily, she relived the scene on the plant floor a few nights earlier when Haney had talked with workers about the men in moon suits, testing machines for spores while employees watched unprotected. Surely, she had thought, postal officials would not allow Brentwood to operate if anyone showed signs of illness. Now she was not so sure. The plant, she told herself, had remained open in the face of danger for one reason. Money.

It had been less than twenty-four hours since Rich called home to tell her that the Brentwood nurse was sending him to their Kaiser Permanente clinic. His primary care doctor, Michael Nguyen, finding nothing visibly abnormal, told him he probably had the flu or a bad chest cold. But following an instinct, the doctor had alerted the Prince William County Health Department that he had a sick patient who worked at Brentwood. By then, health care workers across the nation had been instructed by the CDC to be on the lookout for any suspicious illnesses that could be linked to anthrax. Still, the CDC experts insisted that the chance of exposure in Brentwood was slim since anthrax could not escape a sealed letter.

Nguyen looked Rich straight in the eye and ordered him to get to Inova Fairfax Hospital as quickly as possible. If he was dealing with anthrax, the doctor knew it was a race against time. Nguyen called the hospital in Falls Church to say he was sending a patient for emergency treatment.

While Susan drove Rich to the hospital his condition worsened. He felt awful, his chest tight and his breathing measured. By the time they arrived at the emergency room, he had a throbbing headache and a low-grade fever.

Again, the emergency-room doctors saw a deceptively normal-looking patient. "This guy could have been sitting next to you in a restaurant and you wouldn't have noticed him," said Dr. Thom Mayer, chairman of the hospital's emergency department.[104] But his attending physician, Dr. Cecele Murphy, listened closely to Rich's story. He had never been sick before, he told her, and now he could hardly draw a breath. He was coughing up phlegm. His heart rate and temperature were slightly elevated. The doctor ordered a blood test and a computed tomography (CT) scan to look inside his congested chest.

The CT scan results alarmed the physicians. His lungs had taken on an abnormal shape, divided by a widening gap. "His chest was literally dissolving on him," recalled Mayer. "It's called hemorrhagic necrosis, and what this means is those lymph nodes were blowing apart . . . and you're bleeding into that area."

Murphy informed the D.C. Health Department, then called a

CDC hotline. She found authorities skeptical of her anthrax diagnosis, advising her to treat the patient as if he had pneumonia.

When she was convinced that Rich had anthrax, she went into the intensive care unit to give the news to Rich and his wife. She knew the grim mortality statistics for inhalation anthrax, but also believed that positive thinking, especially in a motivated patient like Rich, could beat the odds. She tried to bolster their hopes, half believing the words herself.

"He's gonna be fine," she said. "We caught it early enough."[105]

She ordered a cocktail of powerful antibiotics, then explained they would begin plasmapheresis, a procedure that would filter out his toxin-laden blood and replace it with healthy plasma. The anthrax poisons were eating his red blood cells, she said, and until they were gone, his health would not improve. She told him they would insert a device in his chest with three tubes—one to remove his blood, another to replace it with plasma, and the third to take blood samples. The device was called a Quentin catheter, which made Rich think longingly of his son, waiting for him at home.

Later that night, Rich came out of the bathroom holding a urine sample streaked with blood.

"How long has it been like this?" the doctor asked, her voice betraying concern.

"Oh, it hasn't been," Rich replied dismissively.

Susan gave him a withering look. It was just like Rich to withhold negative information, to try to appear healthier than he was. "It's been like that for a few days!" she corrected him.

Despite the doctor's hopeful assessment of Rich's condition, Susan reduced the long explanation to its essence: victims of inhalation anthrax had at best a fifteen percent chance of survival, as the only twentieth-century cases had shown. *Everybody dies,* she told herself.

Susan called her mother, who was staying with them in their Stafford dream home.

"I've got to go get Quentin," she said. She wanted their child to have a chance to say good-bye.[106]

She pulled back the curtain surrounding Rich's bed. His breath was more labored by the minute. "Rich? I'll be back in the morning."

Then she headed home. Traffic was calm again at that predawn hour, the High Occupancy Vehicle lanes clear, just as they had been when Rich left for work less than twenty-four hours earlier.

In his dreams, Rich was nine years old again, swimming at the neighborhood pool as dusk set in. A bully swam toward him and overpowered his frail body, forcing his head beneath the water. His brothers saw him thrashing beneath the surface and rescued him from drowning.

Then the nurse would be by Rich's side, fiddling with the IV in his neck, forcing him to return from childhood memories to a small room with loud machines and faded blue curtains that offered little privacy. A minute later, he would be back in Newport News, near the projects where he grew up. Often, he would awake from terrifying dreams, certain he was going to die.[107]

For days, Rich moved in and out of consciousness, only vaguely aware of the constant activity around him. Visitors came and went, and he never noticed.

As poisoned fluids were drained from his body, he shivered under a pile of blankets. "My teeth would chatter," he said. Doctors and nurses surrounded him, poking and prodding, checking the catheter.

He had plenty of time to reminisce. He thought of his declining father, once a deacon in the Baptist church who spent most of his time "making babies." His father would preside over their large family as it spread out along an entire church pew every Sunday morning, with a string of fidgeting and hungry children, anxious for the fried chicken dinner awaiting them back home. But as soon as Rich picked up his high-school diploma, his father had cut the cord.

"Do you want to work in the yard?" he had asked, pointing toward the booming shipyard. "We don't have the money to send you to college. It's time for you to leave."[108]

His mother steered him to Washington in the heyday of federal affirmative action, with agencies rushing to make up for decades of discrimination.

"I took an exam in D.C. for the post office, and I passed it. I'd also taken an exam for the State Department. They said, 'Mr. Richmond, you have the job if you want it.' I was going to work for the South

American division of the State Department, and they were going to send me to school to learn the language," he said.

A day before he was to join the diplomatic service, a USPS recruiter made an irresistible offer: a job paying $1.08 more an hour than the State Department. He grabbed the opportunity.

He started work at the old mail distribution center on North Capitol Street, Brentwood's predecessor. He had never forgotten the stress of that day, and it came back in his fitful sleep.

"It was like hell breaking loose. You didn't stay on one assignment longer than ten minutes. You were herded around like cattle," he recalled.

Zip-code school demanded quick, flawless memorization. "Because I was dealing basically with airmail, I had to learn all the areas of the zip codes of New York," he said. He studied for three months, but when the time came to take the exam, his bosses abruptly moved him to carrier detail.

For almost a year, Rich carried mail through the streets of Maryland and Virginia. His supervisors would lay out a route, drop him off, and give him bus tokens to get back to the office.

"I learned how to carry the mail, and they said, 'No, now we want you to learn Friendship Heights,'" an upscale Washington, D.C., neighborhood. He went back to school for thirty more days, only to get a new assignment soon after.

Brentwood pounded twenty-four hours a day with the sounds of thirty churning machines, each manned by four workers. When one employee took a break, another took over, feeding thousands of letters into equipment that made the whole building smell of warm paper. The noise blended into a smooth cacophony, and then a warning bell would jangle.

"Move over there!" a supervisor would bark.

Rich had made it thirty-four years, finally pulling in $13 an hour and an extra $300 each week in overtime. After all this, it was telling, he thought, that he should contract anthrax not doing his regular job but following instructions to leave his post and clean up behind Joseph Curseen's machine.

He remembered seeing the cleaning man approach Machine No. 17, wearing a face mask and carrying pressurized air. In keeping with the regular routine, the man opened the nozzle and sprayed.

The air went everywhere. A hot blast punched Rich in the face. Later, the Brentwood managers concluded that the cleaning had sprayed anthrax spores throughout the building and caused them to settle deep in Rich's lungs.

From his hospital bed, the whirring sounds of Brentwood's machinery had been replaced by clanking noises from construction outside his room. When Rich had checked in, he peered out his window at an eight-story parking tower surrounded by demolition equipment.

The next time he mustered the strength to look outside, the parking garage was gone.

When Susan led Quentin by the hand to visit Rich Saturday morning around 9:30 A.M., she was stopped at the door.

"The FBI's in there," a nurse cautioned, partially blocking the doorway.[109]

"*You* need to move out of my way!" Susan growled, pushing her way inside. Two agents from the Washington field office sat by Rich's bedside. The Brentwood facility, they told him, was now part of a criminal investigation, and they needed facts: Where did he think he came into contact with the anthrax? With whom did he work? What was his work history? She listened with building irritation as she watched her husband gasping for breath, struggling to respond.

"He can't answer you. He can't even talk," she finally exploded.

Barely audibly he whispered, "It's okay dear," as he waved his hand to calm her.

The others who came to interview Rich were no more sympathetic. The CDC doctors asked so many questions about his health history and the sudden onset of the illness that Rich felt as if they were trying to disprove his anthrax diagnosis. His grown daughter, Alicia, who practically lived by his hospital bedside, finally ordered the feds out of the room so he could get some rest.

Susan had been afraid to allow Quentin to see his ailing father among the tubes and machines, but the child was just happy to see his dad. Leroy was the one who broke down. "He was like, 'Oh my God, my baby's seeing me sick!'" Susan recalled.

Throughout the morning, Susan persisted in her efforts to alert Brentwood to the dangers employees were facing. Finally, she reached one of Rich's supervisors.

"You need to get out of there!" she almost screamed. "The building's contaminated, and you all need to shut it down!"

The building stayed open. She got through to a girlfriend from Brentwood, who spread the word.

Still, the production lines churned, the warning dismissed. Someone scoffed that Susie Richmond, a believer in long coffee breaks and chatty lunch hours, was just trying to find a way to stay home from work.

Susan was livid. "If I just want to be home, why would I lie and say he's got inhalation anthrax?"

By Sunday, postal officials shut down the plant and workers lined up for free antibiotics. Susan was among them. Similar lines had formed in New Jersey near the Hamilton plant, although workers there had grown increasingly distrustful of anything the government doctors or postal management told them to do. When doctors decided to switch the recommended antibiotic from Cipro to Doxycycline because of its lower cost and reduced side effects, incensed Hamilton workers complained that they were being used as "guinea pigs" in a government experiment that was providing them with a "second-class drug."[110]

Back at the hospital, Rich's health declined rapidly. His skin turned grayish, and he had grown progressively weaker. His chest was filled with bloody mucus, which he often spat into a tissue with a deep racking cough.

On Monday, as Susan made her way back to visit him, a Brentwood friend called her to express condolences. The television news had reported the death of an unnamed Brentwood worker; many assumed it was Rich.

Susan broke down and wept, then called the hospital and asked to speak to Carla Asper, an intensive care nurse. The doctors had given her a code word, *cake,* which she could use to get immediate attention.

"This is Mrs. Richmond, and the code word is *cake,*" she said, trying to control her sobs. "How's my baby doing?"

The nurse reported that he was fine. The television account had been about the death of Brentwood coworker Mo Morris of Suitland, Maryland, who had fallen ill over the weekend. Meanwhile, in southern Maryland, the family of Joseph Curseen, the Brentwood Bible-study leader, had rushed to his hospital bedside. Machine No. 17, the same piece of equipment that left his friend Rich struggling for life, had infected Curseen.

Like Rich, Mo Morris ranked as a USPS "lifer," with thirty-three years on the job. He worked as a government mail distribution clerk, inspecting by hand the huge piles of mail bound for government offices, including the Senate office buildings on Capitol Hill. He and Rich often stopped to exchange pleasantries when their paths crossed during shift changes.

Mo felt ill and made his first trip to Kaiser Permanente's Marlow Heights, Maryland, clinic on October 18, a day before Rich sought medical treatment at the HMO's Woodbridge site. Mo, at fifty-five, had a history of hypertension and diabetes. A nurse practitioner at the busy clinic listened to his familiar complaints—a 102-degree fever, persistent aches, difficulty breathing—and diagnosed a virus. Mo was coughing up green phlegm, suggesting a lung infection. The nurse took a culture, recommended Tylenol, and sent him home to rest.

Mo couldn't shake the memory of an incident at Brentwood a few days earlier when a female coworker noticed white powder gushing from an envelope. He had asked supervisors if the material could be anthrax, but no one took it seriously.

After visiting the clinic, he rested at home, but the Tylenol had no effect. At about 6 A.M. on October 21, his breathing increasingly distressed, he dialed 911.

"*My name is Thomas L. Morris Jr.,*" he said. "*I suspect that I might have been exposed to anthrax.*"[111]

He told the operator where he worked and when he thought he was exposed. He said the clinic diagnosed a virus and gave him Tylenol for aches. Now his fatigue was so extreme he nearly passed out every time he walked. He'd started vomiting. The operator dispatched an ambulance.

Morris faded rapidly in the hospital. Within twelve hours, his heart stopped, and efforts to resuscitate him failed. The medical examiner ruled the death a homicide.

Joe Curseen, a second-generation postal worker, at first thought he had a cold, then suspected food poisoning. He worked a full shift that weekend, despite a fainting episode he had in church. Curseen, forty-seven, told his wife he was sure he would feel better.[112]

When he collapsed again, an ambulance raced him to the Southern Maryland Hospital Center. He died on October 22.

Within hours, the FBI launched its investigation of the deaths. Curseen's neighbors watched FBI agents knock on the door, then slip discreetly into the backyard to put on protective suits. From the house, they retrieved the clothing that Curseen had worn on his last day at work and sealed it tightly in plastic bags.

Rich knew none of this as he slipped in and out of consciousness in the Fairfax hospital. In the same intensive care unit, another Brentwood worker, who wanted his identity protected, also fought for his life after calling Kaiser's medical advice line on October 20, complaining of headache and fever. The doctors, suspecting meningitis, did a spinal tap. As with Bob Stevens in Florida, the cultures came back loaded with *B. anthracis.* The patient eventually recovered.

Rich continued to fight against the bacteria that was wreaking havoc on his system. Susan moved into a motel near the hospital so that she could be close to her husband. At night, she would doze in a chair by his bed. Sometimes roused by his tortured breathing or the sound of him struggling to go to the bathroom, she would bolt awake.

"You all right?"[113]

"Yeah. I can't sleep."

"That's all right. Me neither. Do the best you can."

One night, he got tangled in his intravenous cords as he rose from the bed, and couldn't make it to the bathroom in time. He panicked, the first of many such attacks, and cried inconsolably, realizing he had lost control.

"Baby, don't even worry about that. I'll clean that up," she soothed him.

Then Susan noticed blood dripping from the back of his gown and frantically rang for the nurse. She had seen blood before, a bucket of it, spit up from inside his chest as his lymph nodes swelled with infection. This time the problem turned out to be stomach ulcers, developed when the anthrax toxins had seeped into his system.

Sometimes, the nurses would comfort him, sitting by his side as he moaned, "I'm going to die." His lungs would fill with fluid, and the doctors would come to his bedside and tap them with a needle, siphoning the excess liquid one lung at a time into a vacuum container. The first time they drained two liters into the glass bottle. They would repeat the process every few days.

For twenty-seven days he struggled. When it became obvious that he would make it, someone told him about the deaths of Curseen and Morris, and he cried.

One day, Rich awoke to see Postmaster General Potter standing at his bedside, wearing a windbreaker and casual clothes so as not to attract attention. Jaffer, Potter's press secretary, had come with him, but Potter left him behind and scuttled through a back door. Out of respect for the victims, Potter wanted no attention placed on his visit and no photos taken.[114]

He brought Rich a Redskins souvenir and a gigantic get-well card signed by his Brentwood coworkers. Potter stayed only a few minutes, expressing concern for Rich's appearance, although the patient was by then on the upswing.

Rich's doctors finally agreed to let him go home on the condition that he return regularly for plasmapheresis. A barber came to trim his hair and beard, and back in street clothes, he showed off before the kind nurses who had seen him only in a hospital gown.

"We'll see you on *Oprah*!" The nurses laughed, wheeling him to the exit.

Television hostess Oprah Winfrey did, in fact, bring Rich onto the air in May to tell his story. He and Susan found Brentwood's management much less interested. The survivor of the worst occupational disaster in the history of the USPS felt that to his own government he had become like a contagious leper, best ignored, or better yet, forgotten.

United States Postal Service

The Brentwood mail distribution center teemed with life as workers processed three million pieces of mail a day.

Anne Ford Doyle, Inova Health System

Leroy Richmond, his wife, Susan, and his son, Quentin, at the blood donor center where Richmond spoke to volunteers.

Brentwood mail processing facility

Thomas L. Morris Jr.,
March 2, 1946–October 21, 2001

Brentwood mail processing facility

Joseph P. Curseen Jr.,
September 6, 1954–October 22, 2001

New York Post

New York Post staffer Johanna Huden tried not to notice the ugly lesion on her hand until cases elsewhere made her realize she might have skin anthrax.

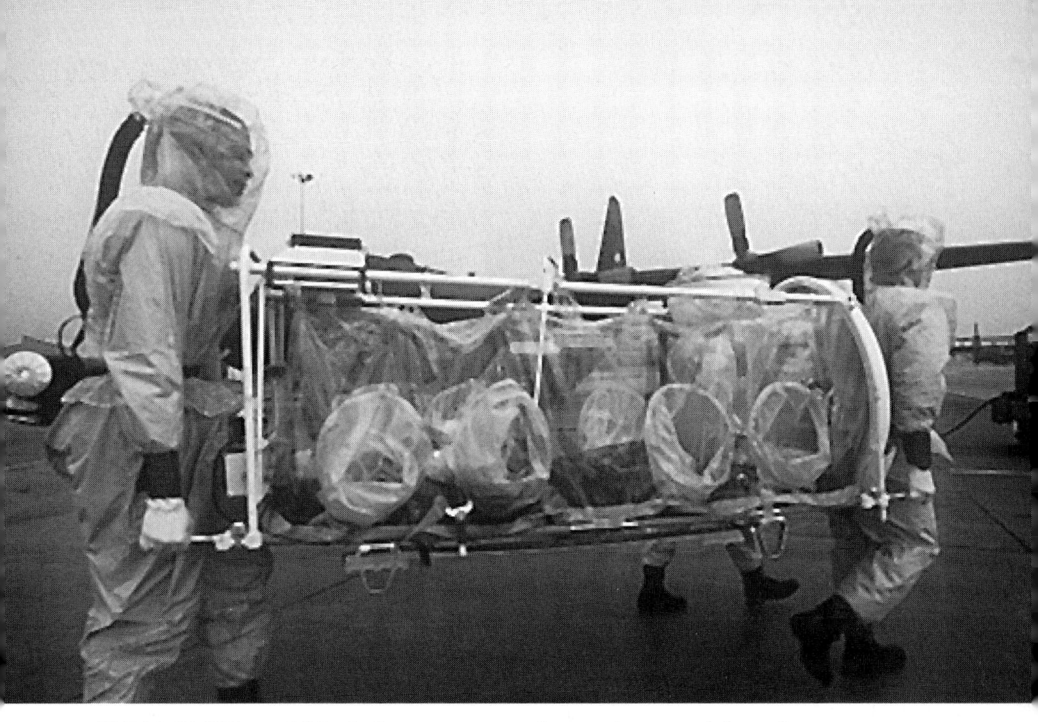

USAMRIID rapid isolation team practices using special equipment to evacuate victims in case of biological attack.
(USAMRIID)

Centers for Disease Control

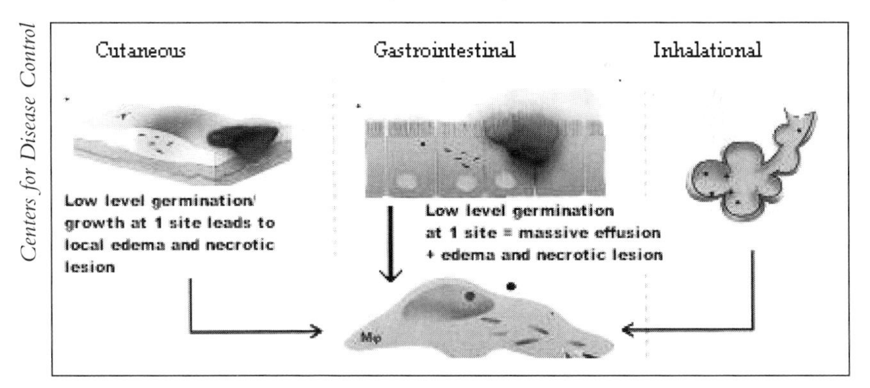

The three types of anthrax infection.

USAMRIID

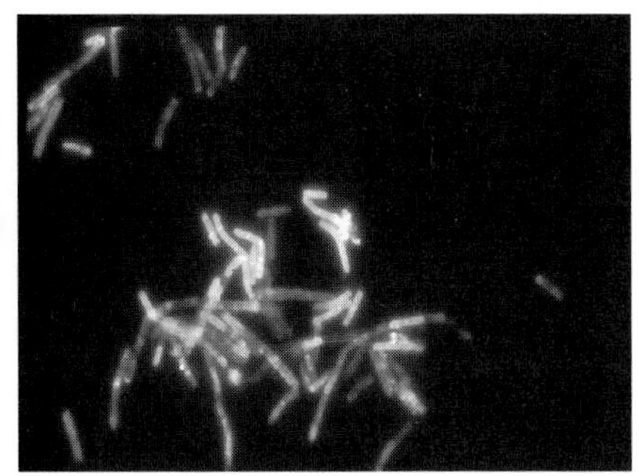

Fluorescent antibody staining of *Bacillus anthracis* vegetative cells as viewed under a fluorescent microscope.

Centers for Disease Control

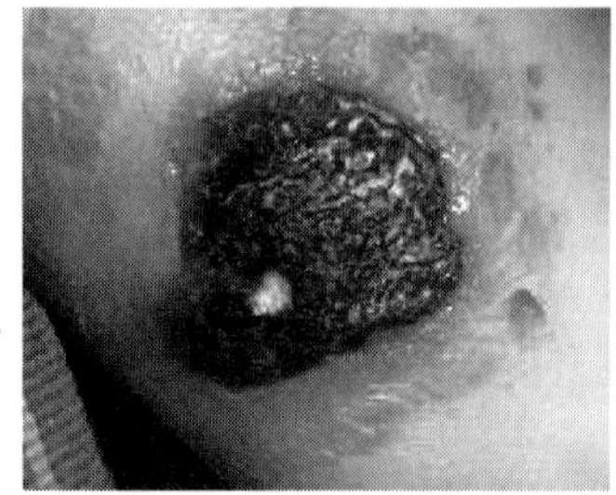

A black scabbed lesion is a symptom of cutaneous or skin anthrax infection.

Dr. Thom A. Myer

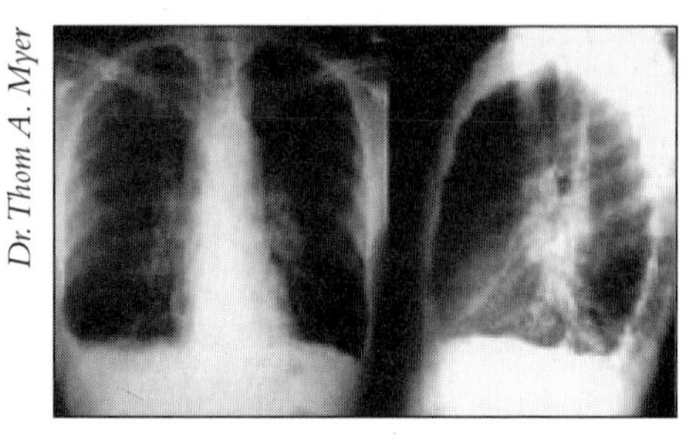

Leroy Richmond's initial radiograph showing a distinctive symptom of inhalation anthrax, a widening of the space between the lungs.

FBI file pho

FBI agents in November, 2001, tested barrels of mail quarantined from Capitol Hill. After one barrel proved to be red hot, agents found an anthrax-filled letter addressed to Senator Patrick Leahy (D-Vt.).

Centers for Disease Control

President Bush shakes hands with Tanja Popovic, director of the CDC's anthrax laboratory, during his November 8, 2001, visit to the Atlanta health agency's headquarters. He was the first sitting president to visit the CDC.

Sen. Leahy's office

Senator Leahy was concerned that the FBI was not being aggressive enough in pursuing the anthrax case.

FBI file photo

09-11-01
YOU CAN NOT STOP US.
WE HAVE THIS ANTHRAX.
YOU DIE NOW.
ARE YOU AFRAID?
DEATH TO AMERICA.
DEATH TO ISRAEL.
ALLAH IS GREAT.

4TH GRADE
GREENDALE SCHOOL
FRANKLIN PARK NJ 08852

SENATOR LEAHY
433 RUSSELL SENATE OFFICE
BUILDING
WASHINGTON D.C. 20510-4502

4TH GRADE
GREENDALE SCHOOL
FRANKLIN PARK NJ 08852

SENATOR DASCHLE
509 HART SENATE OFFICE
BUILDING
WASHINGTON D.C. 2051

The letters mailed to Senator Tom Daschle and Senator Patrick Leahy went through Washington D.C.'s Brentwood mail processing facility, infecting first postal workers, then Capitol Hill staffers. John Ezzell opened both letters and found the highly refined material inside to be almost identical.

Marilyn W. Thompson

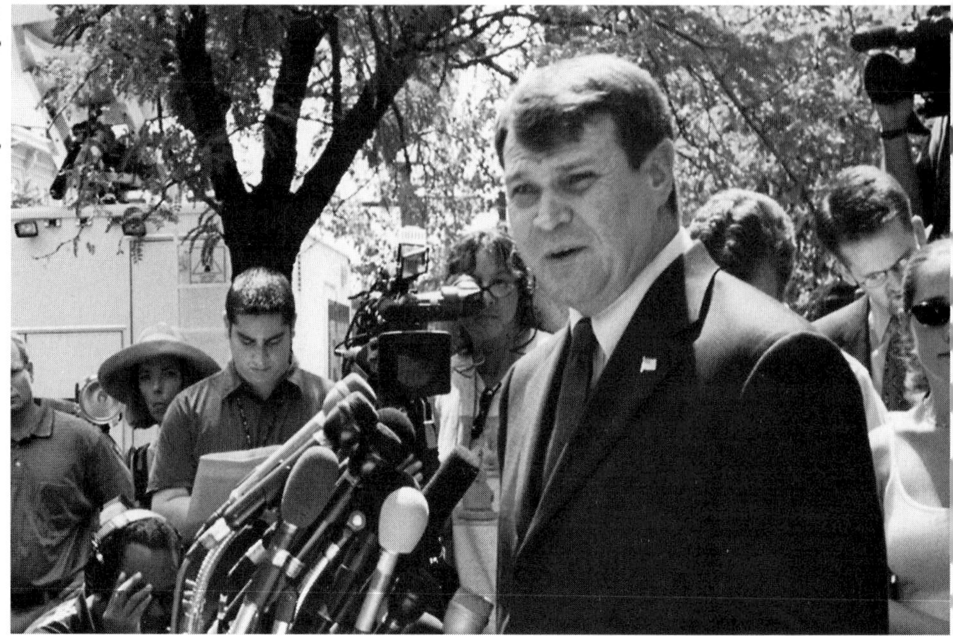

Under a scorching August sun, Dr. Steven Hatfill chastises Attorney General John Ashcroft for labeling him a "person of interest" in the anthrax investigation.

U.S. Army

Researchers working at Fort Detrick's one-million-liter Eight Ball, which allowed scientists a place to safely explode and test germ warfare agents. The structure is the largest aerobiology chamber ever constructed.

FBI file photo

FBI agents took pictures of John Ezzell and his USAMRIID colleagues opening the anthrax letters.

USAMRIID

John Ezzell, director of USAMRIID's Special Pathogens Division, opened the Daschle letter to find anthrax spores of a "weaponized" quality. A few hours later, his assessment had been passed on to the president.

William C. Patrick III

William C. Patrick III, the creator of several patented methods used at Fort Detrick to turn germs into agents of war, estimates that no more than fifty scientists have hands-on knowledge of the weaponization process.

Washington Post

Dr. Jeffrey Koplan, who put off retirement from the CDC to lead the agency through a national emergency, at a House hearing on proposals to combat bioterrorism.

Dr. Stephen Ostroff, who helped manage the anthrax investigation in New York City and the dreaded testing along the city's subway platforms, talks to the press.

AP

AP

Members of the U.S. Marine Corps' Chemical-Biological Incident Response Team scour Capitol Hill for anthrax.

The Aftermath

CHAPTER 12

Patterns and Puzzles

October 22, 2001
WASHINGTON, D.C.

In the days following Richmond's diagnosis and the deaths of two Brentwood workers, something resembling a Greek tragedy was elaborately staged at the highest levels of official Washington. In the grave tones of a commander leading weary troops into yet another battle, President Bush lauded the postal workers who had been stricken in the line of duty and vowed to bring to justice the anonymous murderer who had turned an American postal center into a death trap. A chorus of somber agency heads stepped onto a White House stage to add to the rhetoric and deftly dodged questions about their own roles in an incident that, in many respects, was a preventable industrial accident exacerbated by a series of government blunders and bad judgments.

After finally shutting down Brentwood on October 21, a day after Susan Richmond's desperate phone calls, Postmaster General Potter attended two funerals in suburban Maryland, one of them marked by a procession of red, white, and blue postal delivery vans. Then Potter, Homeland Security Chief Ridge, and members of Congress from Maryland and the District of Columbia praised the late Joseph Curseen and Thomas Morris during a crowded memorial service at a Washington church.

"Joseph and Thomas were perfect examples of what makes the Postal Service great: our people," Potter told the group.[115] Ridge noted that the two men "never formally enlisted in the fight for this war on terrorism, but they found themselves on the front lines. They had no warning that they would give their lives for this cause, but give them they did."[116] Colleagues described Curseen's routine ten-hour workdays and praised Morris's honesty and dependability.

Behind the scenes, harsher words had been flying. No one wanted to take the blame for two deaths that might have been avoided with better medical care and faster response. Although many agencies had played a role, perhaps the worst mistake had been in the Bush administration's reluctance to acknowledge John Ezzell's assessment of the anthrax packed into the Daschle letter. By indulging in a semantic argument about whether the material was truly in a "weaponized" form, officials missed Ezzell's primal scream: the sophisticated anthrax was capable of scattering anywhere, of seeping into human lungs, of killing anyone unlucky enough to venture unprotected into its path.

In public comments, top USPS officials stopped just short of pointing an accusatory finger at the CDC, whose carefully reasoned advice had delayed the distribution of antibiotics that might have saved both workers' lives. Even after Rich was hospitalized, said Deborah K. Willhite, USPS senior vice president for legislative affairs, the CDC and the D.C. Health Department told her there was no reason to close Brentwood or to begin wide distribution of antibiotics. The doctors wanted to wait for proof that Rich had anthrax, and moved forward only when his final round of test results came back positive.[117]

Discussions between the USPS and the CDC had grown increasingly strained in the days before the Brentwood disaster. As soon as she heard that anthrax had been traced from the Hart Building to the Senate mailroom in the adjacent Dirksen Building, Willhite became concerned about Brentwood. If anthrax had leaked outside the letter, postal officials knew Brentwood workers could be in danger. They fretted over the pounding each envelope took from the high-speed sorting equipment.

"Throughout that week, the postmaster general, the chief operating officer, and all of the deputies were talking to anybody that we could . . . urging them to make a recommendation to put [workers] on antibiotics," Willhite said. "Everyone was beginning to have a sense of unease about this, and we felt we had a responsibility. We felt like we needed to reassure our employees." But by the same token, "We're not doctors. We don't write prescriptions."

Postal officials relied on the experts at the CDC. The agency's doctors did what doctors invariably do—offered an informed opinion, not a final judgment, and the CDC maintained that the risk of exposure from a sealed envelope was low.

There were other factors underlying these pointed discussions. From the beginning, CDC officials had offered a surefire way to prevent postal workers from contracting illnesses linked to contaminated mail: simply shut down suspect postal facilities and leave them closed until thorough testing was completed and worker safety guaranteed.

But the idea of shutting down post offices was unacceptable to the USPS and had been strongly discouraged by the Bush White House, which considered the postal system part of the nation's critical infrastructure. As Willhite explained, "Shutting down the United States Postal Service—the system—would have an incredible impact on the daily life of America, economically and socially." The postal system is "the invisible mover of the economy and of the communications that people take for granted. It would have a huge impact for a day's worth of mail not to go through."[118]

In an early conflict, the CDC's Brad Perkins wanted to shut the Boca Raton post office for thorough testing. Postal officials agreed only to stop work overnight, with full operations resuming the next morning. According to Willhite, at no time in the crisis would USPS have agreed to close down the entire system even for a day. The Bush administration solidly backed the postal service's determination to continue delivering the mail.

CDC Director Koplan said that Potter and other officials made clear that closing the postal system—even a region of it—was not a reasonable option. "I remember them saying, 'We're not going to shut down the facilities unless we have to or there is striking evidence.'"[119]

The burden of proof rested with the CDC. "If we wanted to cover our own asses," Koplan said later, "our recommendation to the postal service would be, 'Close it from top to bottom. Let's just close the whole thing down.' That's irresponsible at some level." But shutting down the postal system, he said, would have shielded the CDC from accusations of "doing anything wrong, except for overreacting."

On October 22, news of the Brentwood deaths increased the magnitude of the crisis. A day earlier, postal officials had shut down the 500,000-square-foot plant, quickly shifting to an emergency backup plan for handling Brentwood's incoming mail. Millions of pieces of mail still inside the plant became caught in limbo while officials decided how to go about decontaminating them.

That morning, President Bush greeted postal and union representatives at the White House, telling them the USPS had been drafted into "an unprecedented war" against terrorism. Potter, the onetime mail clerk, now found himself involved with CIA chief George Tenet and FBI chief Mueller in meetings on terrorism policy. After meeting with Bush, Ridge, and the CDC's Cohen, Potter took questions from the press.

When asked whether mail to the Washington, D.C., area would be halted until the contamination could be contained, Potter was steadfast.

"No, we don't intend to curtail mail delivery. We're not going to be defeated. . . . We have no intent to stop delivery of the mail, unless we have a situation where people—where we suspect anthrax, and obviously then we'll pull back."[120]

Ridge stressed that "aggressive and proactive treatment" was under way to protect employees at Brentwood and the Baltimore-Washington International Airport postal facility where Rich had worked part of each day. "We took immediate steps to treat each worker who might have been exposed."

Cohen defended the CDC's reluctance to dole out antibiotics without proven need. "There is risk in prophylaxis when it is not necessary. One of our basic goals is to identify who is at risk," he said. Using antibiotics unnecessarily could cause serious problems, he said.

Potter expressed his regret over the two deaths and his hope that the tragedy would strengthen the commitment of the postal employees to keep "moving America's mail."

"We will overcome this," agreed Vince Sombretto, president of the National Association of Letter Carriers, standing at Potter's side. "We will not be deterred from doing our job."

Again, the White House sought to project an image of control, downplaying the dangers and reiterating to the public that the damage had been contained.

Only a very few suspected that the nightmare was not over. One of them was Rima Khabbaz, the CDC doctor who had become, in many press accounts, an easy scapegoat for the Brentwood blunders.

When hospital doctors treating Leroy Richmond first reported that they suspected inhalation anthrax, Rima Khabbaz had been so confident that the diagnosis was wrong that she sent her team into Brentwood without protective clothing. "Risk, in our minds, did not exist," she recalled.[121]

She had been holding around-the-clock consultations with the CDC anthrax experts, all of whom agreed that the risk of exposure to inhalation anthrax inside Brentwood was low, based on the agency's recent experience with mail facilities in Florida and New York. At that point, the only mail worker known to be infected was Teresa Heller, the New Jersey carrier. Khabbaz assumed that if a Brentwood worker had handled the Daschle letter, the worst possibility was a case of cutaneous anthrax, which could be diagnosed early and treated with antibiotics.

Then, on the night of October 19, Khabbaz learned about Leroy Richmond's case. She set up her conference call with her Atlanta colleagues the next day to ask if she could begin distributing antibiotics, and if so, to which workers.

Someone asked her to explain the setup of the plant floor. Khabbaz described the open plant, the absence of dividers and barriers, and estimated that two thousand people worked in the room.

She paused, waiting for a reaction. The voices were all silent. The scale of the potential exposure was only beginning to dawn on federal doctors as they made arrangements for another massive distribution of antibiotics.

On Saturday night October 20, before Richmond's diagnosis was official, Khabbaz and her team tried to track down those who worked near Richmond. She wanted them notified and pulled to the front of the antibiotic line. Guided by plant manager Haney, her team also began contacting local hospitals and clinics in Maryland and Virginia and made phone calls to anyone at Brentwood who had missed a day of work that week. Neither of these efforts put them in touch with Curseen or Morris, who had both worked a full week, then retreated to their homes to struggle with worsening illness.

On Sunday, October 21, the CDC officials confirmed Rich's diagnosis and recommended prophylaxis for all Brentwood workers. Khabbaz was relieved when the plant was finally shut down later that day. The surveillance effort in the hospitals and clinics seemed to be going well. Antibiotics were flowing freely, with enough to treat 2,400 people.

But when Curseen and Morris showed up in emergency rooms in rapid-fire succession, both dying quickly, Khabbaz threw out all of her assumptions.

From Atlanta, Brad Perkins tried to convince Khabbaz that the agency's decisions in the Brentwood case were defensible. "You could take a more conservative position and say, 'Well, we should have closed all the post offices,'" he said later. "But that would have been an extremely difficult position for the government to take without more definite information about risk."[122]

On a diagram of the plant floor, Khabbaz could see the hot spots where the four affected Brentwood workers had worked, and the locations only confused matters. Two workers had been stationed near Machine No. 17 at one end of the plant; the others had worked at the opposite end in the government-mail-sorting area, where letters were brought over in trays and inspected by hand. Two of the four had not even been in the plant at the time the Daschle letter passed through Machine No. 17.

Surface wipe results, available on Tuesday, October 23, from around the plant floor showed anthrax contamination at both ends of the plant, yet other parts of the floor had tested negative. So had the customer service area, the second-floor administrative offices, the loading areas, and the express mailroom, where Richmond normally worked.

Although she had studied many dangerous illnesses, Khabbaz was trained as an epidemiologist, not as an anthrax expert, and had found many aspects of the Washington cases baffling. Experts once believed that anthrax was unable to reaerosolize once it had landed on a surface. But in D.C., the germs seemed to have spread through the air at different times, turning up in the most unexpected places. At a State Department mail annex in Winchester, Virginia, an hour's drive away, spores infected veteran worker David Hose with inhalation anthrax on October 25. Khabbaz and her colleagues had begun to feel increasingly frustrated.

New Jersey's problems also had multiplied. After Hamilton USPS worker Norma Wallace came down with inhalation anthrax, cutaneous cases had surfaced, both in the mail system and among the general public. A bookkeeper at a Hamilton accounting firm developed an anthrax lesion after handling mail delivered to the office. Concern heightened when anthrax was found at four of the forty-nine mail facilities that fed into the Hamilton processing center.

Over and over, Khabbaz relived the decision to delay distributing antibiotics to Brentwood workers—an action that had seemed so well-grounded, yet looked tragically wrong after the loss of two lives. She agonized as she thought of Leroy Richmond, asking herself, "Aerosol. How far? So many people. Why him?"[123]

From Washington, Khabbaz tried to defend the CDC actions to the public, taking the brunt of much of the criticism.

Koplan, in Atlanta, felt there was no reason Khabbaz should feel solely responsible. "I think there was kind of unanimity amongst those of us who were epidemiologists in public health, which was that . . . the letter went through a large number of places. There was as yet no evidence of contamination; you had to go look and find it first. And there was reluctance to put what would have been many thousands more people on prophylaxis with no sense of a risk to them at the time. That's what the thinking was. So it wasn't just Rima." [124]

While Khabbaz was studying the Brentwood sampling results, she had an epiphany. The anthrax victims from Brentwood worked in different areas of the building. What was the likelihood that they had all been exposed from the same source?

"I had this terrible feeling in my stomach; [I] actually cringed," she recalled. "Could there be another letter?"[125]

The first inexplicable case surfaced in New York. The victim was a woman so anonymous that investigators struggled just to piece together her life, much less find a cause for her disease.

A Vietnamese refugee who had become an American citizen, sixty-one-year-old Kathy Nguyen lived alone in a Bronx apartment and commuted by subway to her job in the stockroom of the Manhattan Eye, Ear and Throat Hospital on the Upper East Side. A diligent employee, she stocked operating and recovery rooms with medical supplies. Her job did not entail handling hospital mail.

Neighbors and acquaintances knew little about her personal life. She was divorced; her son had moved to Europe with his father and died in a car accident. Her life seemed to consist of a single routine: going to work and returning each evening to the solitude of her apartment.

Daily, she commuted on foot along an access road to the Sheridan Expressway, past a car wash, a mechanic's shop, and a few other neighborhood businesses. After the block-and-a-half walk, she would climb the stairs to the elevated Whitlock Avenue station, then ride the No. 6 train past thirteen or fourteen stops to the East Eighty-sixth Street station and walk the rest of the way to the hospital.

She first complained of muscle aches and fatigue at work on October 25. Over the next two days, the symptoms came to include chills, breathing problems, and a nasty cough that brought up traces of blood.

On October 28, after a full day of work, she asked her landlord to drive her to the emergency room. He took her to nearby Lenox Hill Hospital. Nguyen was conscious and alert and had no fever upon admission, but her breathing was labored, necessitating an oxygen mask. Doctors tentatively diagnosed pneumonia and began administering antibiotics, but her condition rapidly deteriorated. Early on October 29, doctors gave her a breathing tube and moved her into intensive care. An emergency CT scan found thickened mucus and massive bleeding in her lungs. Doctors drained 2.5 liters of fluid from her right lung and 1 liter from the left. They alerted city health officials, who, in turn, called the CDC.

The CDC's Stephen Ostroff was in New Jersey to meet with concerned postal workers in Hamilton when the Atlanta office called with the news about Nguyen. He was told that there was a woman bleeding into the middle of her chest who might have anthrax, a ruptured aorta, or another diagnosis. "As soon as I heard it, I knew it was going to be another case of anthrax. The woman bore no similarities to the other anthrax cases. And we all had this terrible sinking feeling that this was the next phase of the episode. She didn't work at ABC, CBS, Fox, whatever, and she didn't look, sound, or feel like the other patients. Plus, she had inhalational disease."[126]

The next phase, as Ostroff defined it, involved cross-contaminated mail—letters that had passed through contaminated mail facilities and made their way out into the general population. Ostroff remembered the questions from the press and the postal workers. "They're asking us, 'How can you reassure us that there's not going to be another letter that comes through the system that will kill us all?'" Postal workers at Manhattan's Morgan postal facility had filed a lawsuit seeking closure of the plant after four sorting machines tested positive for anthrax.

When Ostroff arrived in New York, he had a chilling sense of urgency. "If this woman got it, pretty much anyone else in the city could have it," he said. "We were trying to think about how and where this woman was conceivably exposed. By then, we were well into the issue of cross-contaminated mail. So we thought, 'Maybe this is the first instance of an innocent bystander.' We also had to be concerned that maybe someone did do some sort of release where she happened to be standing."

Nguyen died on October 31. Since the cause of her infection was still unknown, health officials feared the worst: that she might have contracted the germ inside Lenox Hill. In consultation with Mayor Giuliani's office, they took the extraordinary step of shutting down the hospital until testing could be completed, moving its small inpatient caseload to other city facilities. Though it was still an undertaking, the officials were grateful that Nguyen had not checked into one of the city's large hospitals, with hundreds of patients who would have to be relocated.

As Ostroff's agents began pursuing the case, following the same murky leads as the New York field office of the FBI, Nguyen emerged as everybody's worst epidemiologic nightmare. Her life had been so private that few people could attest to her movements. She attended Mass every Sunday at St. John Chrysostom Church. She occasionally took the subway to Chinatown to buy Vietnamese ingredients for cooking. She knew the owners of a restaurant on East Eighty-sixth Street. She had an account at a bank but rarely went there, preferring to use money orders to pay her bills. Tests of her apartment yielded no sign of anthrax, and hazardous materials teams reported no spores in her workplace. Investigators could find no evidence of contaminated mail in either location.

In desperation, authorities posted Nguyen's photo on street corners and at subway stops, hoping that new facts would emerge. None did.

"And so the key to the investigation," Koplan said, "without attributing any motive, lack of motive, or theory of what happened—is just knowing where she was. That's the crucial thing. And so [with] all of the posters you see out at churches, and on the streets, and elsewhere and on TV, the intent is [to learn], 'Did someone know her even casually?'"[127]

Mayor Giuliani, City Health Commissioner Neil Cohen, and other baffled city and federal officials had considered the awful possibility that Nguyen had contracted the disease in the city's labyrinthine subway system, which has nearly seven hundred miles of track and 490 stations. With Manhattan still reeling from the infrastructure damage done on 9/11, no one wanted to face the possibility that a terrorist could have descended into the bowels of the city to release anthrax in its most vital transportation system, carrying nearly 7 million passengers on an average weekday.

Ostroff said the subway discussion envisioned a "nightmare scenario." Authorities debated how to do sampling in inconspicuous ways so that riders would not panic. Some argued that perhaps the entire system should be shut down until its safety could be proved. If the system had been contaminated, "how many people were theoretically exposed that we would have to worry about? . . . It just raised a tremendous number of questions."[128]

Finally, the decision makers decided to move ahead. They devised a testing plan that included stations Nguyen frequented, as well as other stations that would serve as "control" points.

On November 9, a brief notice in a press release about a *New Yorker* staffer's negative test results said that as part of the ongoing criminal and public health investigation into the Nguyen case, the city Department of Health would "work with the CDC to conduct testing in subway stations that were frequented by the deceased victim."[129]

Ostroff's sampling team brought back encouraging news. After testing a variety of sites along Nguyen's route, both at elevated stations in the Bronx and busy underground stops along the No. 6 line, the CDC reported negative results. A potential public health emergency appeared not to have materialized, but the mystery of what happened to Kathy Nguyen, Ostroff said, "remains an enigma to us."[130]

On November 21, the death of ninety-four-year-old widow Ottilie Lundgren sent another shock through the nation's public health system. By far the oldest victim, Lundgren rarely left her home in the small town of Oxford, Connecticut, except when friends would drive her to the doctor or the local beauty salon. Caretakers brought in her daily mail, as Lundgren was too frail to walk to the rural box.

Lundgren had been admitted to the hospital complaining of shortness of breath, dehydration, and fever, and doctors initially suspected a urinary tract infection. They drew blood and sent it to the hospital laboratory. Microbiologist Harold Hebb, who had been analyzing samples for thirty years, saw rod-shaped bacteria in the culture, and thinking the test was wrong, drew four more samples. All showed the same results. He contacted an infectious disease specialist to ask a question that soon was echoed in public health agencies across the country: "Where in the hell does a ninety-four-year-old lady in the backwoods of Oxford get a case of anthrax?"[131]

Peg Crowther, a longtime friend of Lundgren and one of several people who regularly checked her mail and took her on errands, was placed on Cipro, as was anybody who visited her. All the CDC tests of Crowther's home and the inside of her car came back negative.[132]

Authorities at first suspected contaminated mail—specifically, a routine mailing from a Connecticut senator or congressman that might have passed through Brentwood during the critical days of mid-October. But they could find no evidence to support their theory.

When an elderly neighbor of Lundgren's was found dead in his home, apparently undiscovered for several days, doctors suspected a connection. One called Sherif Zaki's CDC pathology lab, and Zaki coached him through an autopsy by speakerphone to determine if the man died from anthrax. The results were negative.

Because of the precedent of the case, Zaki sent a CDC pathology team to assist in Lundgren's autopsy.

From Washington, HHS Secretary Thompson, his adviser Henderson, and Anthony Fauci, director of the National Institute for Allergy and Infectious Diseases, appeared with Koplan at a televised briefing to discuss the latest case with reporters. Lundgren was the fifth person to die of anthrax.

Officials hoped that clues would emerge to explain Lundgren and Nguyen's perplexing deaths. But in the absence of real answers, their concern was growing. The anthrax that spewed through the Hamilton and Brentwood postal distribution centers could have tainted huge volumes of mail, hundreds of thousands of pieces, Koplan guessed. And as that mail made its way to distant destinations, officials could only pray that it would not leave other innocent victims infected or dead.

CHAPTER 13

Too Hot to Handle

November 16, 2001
WASHINGTON, D.C.

Of course, there was another letter, just as Rima Khabbaz had feared. It took a while to surface. The envelope had been addressed to Senator Patrick Leahy, the Judiciary Committee chairman. Somewhere along the line the envelope's zip code, 20510, had been incorrectly read as 20520 and the letter was dispatched to the U.S. State Department's Annex 32, a small, independent mail-processing facility in Winchester, Virginia. Almost identical to the Daschle envelope in its crooked script and threatening language, it had come into Brentwood on October 11 or 12, the same time as the Daschle letter, and bore the same October 9, Trenton, New Jersey, postmark.

If the letter had not followed a wayward path, it might well have ended up in Leahy's office on October 15, the same day that the Daschle intern ripped open the deadly envelope, exposing a suite full of staffers. The results, however, would have been much different. Leahy's chief of staff, Luke Albee, had become so alarmed by the anthrax letters in New York and Florida that he consulted with office manager Clara Kircher, and they made a command decision on the day the NBC letter surfaced: no incoming mail would be brought to the office and opened until the threat passed.[133]

In a memo to Leahy, titled "Taking Security in Our Hands," Albee informed the senator that "in light of the latest anthrax news at NBC, WSJ *[Wall Street Journal]* and the state department, Clara and I have decided to stop opening all mail until we can get some guidance from the 'experts' charged with protecting us. . . . It seems quite plausible to me that a member of Congress could be on their mailing list."[134]

A slender, angular man known for his keen wit, Albee suspected that the anthrax mailings were the work of a right-wing zealot deliberately targeting liberals. In the tense days since 9/11, Leahy had been a prominent figure on Capitol Hill and the target of sharp criticism from the Bush administration. Although communication with constituents is the lifeblood of a Senate office, stopping the mail until the threat abated would protect both his boss and the crew of young interns who normally handled the daily barrage of letters. Leahy would later call Albee a hero for his preemptive action, although everyone considered the chief staffer paranoid at the time.[135]

The path of the anthrax-filled Leahy letter is unclear after it arrived at Brentwood. The misread zip code appeared to have sent it to the State Department facility, where it was redirected to the Capitol, its delivery stalled by the disruptions on Capitol Hill after the discovery of the Daschle letter. Then, on October 17, the FBI quarantined all remaining U.S. Capitol mail, stuffed it into 635 plastic garbage bags, and placed the bags into 250 sealed drums. The drums were transported to a General Services Administration warehouse in Springfield, Virginia, which was sealed in plastic to prevent hazardous material from escaping. For a month, the mail sat there awaiting further analysis.

On November 16, a team from the FBI and the EPA, wearing protective suits and respirators, entered the plastic-encased chamber. Working in two-man teams, they opened the drums and began sorting through the bags, a tedious process that went on for seven days. It was the largest hazardous materials operation ever conducted by the FBI. Agents awkwardly maneuvered their gloved hands into each bag to cut a narrow slit that allowed them to swab inside. Six bags tested hot for anthrax, but one was exceptionally high, registering twenty thousand spores in the quick test.[136]

Agents knew they had scored. They opened the bag and sorted through it, letter by letter, until one agent came across an envelope that, like the one addressed to Daschle, bore telltale childish handwriting and was taped shut. Its return address was also the fictitious Greendale school of Franklin Park, New Jersey. What appeared to be anthrax was spilling from the envelope.

The agents sealed the tainted envelope in a series of Ziploc bags and set out for Fort Detrick, where the materials could be analyzed by John Ezzell's team. Unopened and virtually untouched, it was by far the most valuable piece of forensic evidence yet in the burgeoning criminal case.

Patrick Leahy had been heading home to Virginia after a grueling Senate workday when the cellphone in his car rang. FBI Director Mueller needed to speak with him. It was a few days before Thanksgiving, and Leahy had been pushed by the frenetic legislative activity since 9/11—the controversial USA PATRIOT Act, the establishment of a fund for terrorism victims, the usual debates on the appropriations bill—all made more trying by the scare surrounding the Daschle letter. Mueller told Leahy that a hazardous materials team had been proceeding slowly through the quarantined mail in northern Virginia and had reached the last barrel. "Before he even got the words halfway out of his mouth, I kind of figured what's going to happen," Leahy said.[137]

The FBI director told Leahy a letter addressed to him had been packed with anthrax.

Mueller then promised to give the senator a full briefing the next morning at his office. Leahy agreed to release a press statement and inform the Capitol police.

The brief, confident statement that Leahy later gave the news media reflected none of his inward anxiety. He had been threatened with death before, back in his days as a prosecuting attorney, but those bluffs had come from criminals who were not very smart. This time, the killer was obviously intelligent, capable of using advanced techniques and leaving no traces of his or her identity. Even more troubling was the lack of a clear motive.

In his statement, Leahy trusted law enforcement with these unsettling questions, commenting, "I will leave it to the proper authorities to report what they know and the procedures they are taking. I am confident they are taking the appropriate steps and that eventually they will find this person."[138]

The release politely dodged a concern that had been growing in Leahy and others on his committee that the FBI was not being aggressive enough in pursuing the anthrax case. Leahy's confidence in the FBI had sagged as he had watched its errors in recent years, and his committee had pushed for reforms and reorganization. Democrats on the Judiciary Committee had made clear their intention to hold the bureau accountable for its performance in the anthrax matter, as well as missed clues in the 9/11 tragedy.

The first showdown came during a November 6 hearing before the Subcommittee on Technology, Terrorism and Government Information called to delve into laboratory security breaches that might have led to the Ames strain falling into a murderer's hands.

The FBI sent James T. "Tim" Caruso, deputy assistant director in the counterterrorism division, to testify. Only a few days earlier—the day of Ottilie Lundgren's death—the FBI had announced the imminent departure of Deputy Director Thomas J. Pickard. The bureau's second-in-command and Caruso's boss, Pickard had accepted a security job with a pharmaceutical company. His resignation would again break the continuity of the terrorism probes that had stretched the bureau to its limits since 9/11.

Caruso tentatively answered the committee's pointed questions. In a typical exchange, Chairwoman Dianne Feinstein (D-Ca.) asked him how many U.S. labs handled anthrax:

CARUSO: We do not know at this time.
FEINSTEIN: You don't know that?
CARUSO: No, we do not. We're pressing hard to determine . . .
FEINSTEIN: Could you possibly tell me why you do not know that?
CARUSO: The research capabilities of thousands of researchers is something that we're just continuing to run down. I know it's an unsatisfactory answer, and unsatisfying to us as well.[138]

Senator John Edwards, a North Carolina Democrat, joined the grilling. He asked Caruso if the FBI could identify every lab in the country with access to the Ames strain. Caruso said the FBI did not have the ability to do so, that it was "too diverse a population."

EDWARDS: But the bottom line is this: As of now, you don't know where the anthrax came from and you have not been able to identify all the people who may have access to it. Is that fair?

CARUSO: That's correct.

Speaking to reporters afterward, Edwards lamented that the FBI was not any closer to figuring out the answer to these questions.

Three days later, in a show of progress, the FBI released the linguistic and behavioral assessment of the person responsible for the anthrax letters. A handwriting analysis pointed out the author's eccentricities, such as writing *can not* as two words instead of one, as is more common. It noted the exclusive use of uppercase letters and the tilt of the writing on the envelopes. The choice of prestamped envelopes eliminated a potential clue—the human saliva normally used to lick a stamp.[140]

In its profile, the FBI described the assailant's likely characteristics:

The killer likely has taken appropriate steps to protect his own safety, which may include the use of an anthrax vaccine or antibiotics.

He has access to a source of anthrax and possesses the expertise to refine it. The suspect possesses or has access to some laboratory equipment, such as a microscope, glassware and centrifuge.

His thought processes are organized and rational during his criminal behavior.

He is familiar, directly or indirectly, with the Trenton metropolitan area, and is comfortable traveling in and around Trenton. However, he may not currently live in the Trenton area.

The killer did not select his victims randomly. He made an effort to identify the correct address, including zip code, of each victim and used sufficient postage to ensure proper delivery of the letters. The offender deliberately targeted NBC News, the New York Post *and the office of Senator Daschle, and possibly AMI in Florida. These targets are probably very important to the offender and may have been the focus of previous expressions of contempt, which may have been communicated to or observed by others.*

He is a non-confrontational person, at least in his public life. He lacks the interpersonal skills necessary to confront others. He chooses to confront his problems by "long distance," rather than face-to-face. He may hold grudges for a long time, vowing that he will get even with "them" one day. There are probably other, earlier examples of this type of behavior. While these earlier incidents were not anthrax mailings, he may have chosen to anonymously harass other individuals or entities that he perceived as having wronged him. He may also have chosen to utilize the mail on those occasions.

He prefers being by himself more often than not. If he is involved in a personal relationship it will likely be of a self-serving nature.

The bureau identified behavior that the killer may have exhibited after 9/11 but before the mailings:

He may have become mission-oriented in his desire to undertake the mailings. He may have become more secretive and exhibited an unusual pattern of activity. Additionally, he may have displayed a passive disinterest in the events which otherwise captivated the nation. He also may have started taking antibiotics unexpectedly.

The FBI also said that during the course of the anthrax mailings and the media coverage of them, the perpetrator might have exhibited significant changes, including:

Altered physical appearance
Pronounced anxiety
Atypical media interest

Noticeable mood swings
Increasingly withdrawn behavior
An unusual level of preoccupation
Unusual absenteeism from normal activities
Altered sleeping and/or eating habits

The profile concluded by saying that strange behavior would have been the most noticeable after the mailings, the first death of an anthrax victim, media reports of each incident, and the deaths and illnesses of nontargeted individuals.[141]

The FBI's statement immediately came under criticism. Skeptics accused the bureau of too quickly dismissing possible involvement by a foreign terrorist, especially hints that the 9/11 hijackers might have experimented with anthrax. They noted that a similar analysis, done years earlier, of the mysterious Unabomber, speculated that the man was a well-dressed intellectual, not the unkempt recluse holed up in a remote cabin that authorities discovered when they arrested Ted Kaczynski.

The emergence of a new anthrax letter and signs of trace contamination in several other Senate suites touched off a new round of concern on Capitol Hill. The Russell Building was closed for testing. Hill staffers met regularly with doctors, but the information they were given seemed contradictory and confusing.

Leahy tried to put the death threat out of his mind and keep his sense of humor. The day after his FBI briefing, he and his wife attended a Georgetown dinner party. The hosts had heard about the contaminated letter and worried that the subject might be too sensitive to bring up in a social setting.

"I arrived at their house and had my hands behind my back and said, 'I'm sorry I'm a little bit late.' I pulled my hands out in front. I had big bright yellow rubber gloves on," Leahy recalled.

"I said, 'I have a bunch of mail I still haven't gone through. Can I borrow your kitchen table?' That kind of broke the ice and we went on from there."[142]

Because of the purity of the Leahy evidence, John Ezzell's team and its FBI counterparts wanted to proceed carefully in opening the letter to let good science—not public pressure—drive the investigation.

Some of the material had already leaked out and proved to be the same threatening quality as the spores in the Daschle envelope.

"This was our opportunity," Ezzell recalled. "This, to me, was pristine. We could feel the letter; we could tell there was powder inside the envelope."[143] The team wanted to protect both the laboratory workers and the powder itself from contamination.

Ezzell's USAMRIID team spent weeks with the FBI figuring out its strategy. The crew wanted all surfaces in a specially designated lab to be perfectly clean and dry so the material would not be altered.

"Every step of the way . . . had to be investigated," Ezzell said. "And so there were a number of things that we had to work out and, in some cases, practice—how to handle the powder, how to handle air movement. It took a lot of meetings, talking to various companies. We brought in special consultants. At the same time, we were getting pressured by the press. . . . The funniest thing was when a newspaper reported that we were bringing in a robot to open up the letter. And Senator Leahy was also getting anxious for us to open up the letter."

Ezzell's team and the FBI stuck to the plan. No matter what the pressure, the scientists would open the letter only when they were ready. They knew that their work in the laboratory could have enormous importance for the prosecution of the killer when the FBI made an arrest. "If we had not taken our time and we would have screwed this up, how stupid would we have looked? We didn't care how much criticism we were getting. It paid off. We didn't want to take a chance," Ezzell said.

Finally, three weeks later, on December 5, after numerous tests and trial runs, the team decided its preparations were in place. With important lessons learned from opening the Daschle letter, scientists this time proceeded directly to a special room within the BSL3 containment lab, wearing full protection. Two trusted support personnel would back Ezzell: his frequent collaborators Candi Jones and Dr. Jeff Teska.

"We practiced what we were going to do," Ezzell said. "We had every step planned out, and as we proceeded, Jeff documented it, wrote everything down. A special hood was set up in front of a viewing window through which the FBI could monitor the opening of the letter. We had an FBI photographer who set up his equipment. Some FBI and other folks were on the outside watching."

The procedure began around noon. Team members wore protective suits, double gloves, and respirators. Ezzell took the lead, prying open the sealed metal can in which the letter had been placed for protection. The team had devised a technique to determine exactly where the powder had settled inside the envelope so that Ezzell could scoop it out with a micro-spatula and move it into a secure vial. As he had learned from experience, great care had to be used to keep the spores from dispersing into the atmosphere.

"Everybody was very tense. Nothing was going to be dropped," Ezzell recalled.

Working delicately through rubber gloves, he found the powder lumped into a corner and began lifting it out with his scoop. At one point, despite the practice, a minuscule amount of powder slipped off the spoon.

One of the spectators outside the window screamed, "Oh, no!"

"When you're working with the whole world watching you, the last thing you need is to hear through the window someone screaming. I was hooked up to a microphone so I could talk back and forth to Special Agent Scott Stanley outside," Ezzell said.

Without looking up, Ezzell said, "I heard that!" The situation was under control—the powder had fallen into a catch tray and was contained.

The procedure took about four hours. The letter, once extracted, cleaned, and photographed, proved to be a duplicate of the Daschle message. Both the letter and envelope were neutralized and stored securely as evidence.

The anthrax spores from within the envelope, still in their virulent form, were stored in a protective vessel as secure as Fort Knox. The canister was spirited away for storage in an undisclosed location in Fort Detrick. Ezzell became its official custodian.

Every so often, the FBI would call Ezzell and ask him for some of the Leahy material so it could run yet another battery of tests. The bureau had engaged a handful of labs to do cutting-edge analyses to pinpoint the material's age, its chemical composition and additives, and the types of equipment used to dry or refine it.

Ezzell would go to the vessel and fetch a minuscule sample. "We are very stingy with the stuff—we don't want to waste any of it," he said.

He took comfort in knowing that his work might help capture the killer. One night, flying back into Washington, Ezzell looked down from the airplane window. "You get so caught up in the floor and walls of the laboratory, and then you fly over large areas and you look down at all the homes and all the lights down there," he said. "It makes you think how what you're doing is helping protect all these people."

In the days before the Thanksgiving break, the FBI interviewed everyone who worked with Leahy, looking for commonalities that might link Leahy and Daschle in the mind of a calculating murderer.

The USPS had shipped all of the withheld Capitol Hill mail to Ohio for irradiation treatments that would kill any remaining anthrax spores. Months later, the letters and packages came back brittle and reeking of chemicals. Staffers complained of skin irritations, dizziness, and headaches.

When the mail office asked Leahy's staffers to pick up bundles of decontaminated letters, no one wanted to step forward. Clara Kircher went to fetch the mail herself, opened letters in a conference room, and immediately felt sick to her stomach.

Again, the doctors and experts showed up to allay concerns. They were being pressured by members of Congress to direct the federal health agencies to investigate them. The cause of many of them turned out to be simple: the USPS contractor doing the decontamination was using much more radiation to kill the anthrax spores than was needed.

Again, Luke Albee took matters into his own hands. He and Kircher decided that they would personally open the mail until a staffer could be trained to take over the job safely.

They carted the letters into a conference room beside Leahy's office, dumped piles onto a long, polished table, and pushed open the windows. They pulled on rubber gloves and, one by one, went through the letters. All nonessential mail went into a Dumpster. Letters from constituents were set aside for inspection. Then, after carefully examining each envelope for warning signs, Albee, the highest-paid staff member on the personal payroll, did an intern's job. He ripped each letter open and crossed his fingers, hoping that no white powder would emerge.

CHAPTER 14

Taking Stock

February 2002
BOCA RATON, FLORIDA

Ernesto Blanco recovered first. The oldest anthrax survivor went back to work managing the mail at American Media's new temporary location, fifty-three thousand square feet of sterile space in a Boca Raton office park that felt more like an insurance agency than the newsroom of an irreverent tabloid. The contaminated building in which Blanco and his friend Bob Stevens had come into contact with anthrax had been sealed and surrounded by armed guards while AMI wrestled with the EPA over who should bear the cost of decontamination. Initially, the EPA had suggested the building would be eligible for federal cleanup funds. Then, on November 12, the agency declared that it would be up to AMI and its private insurers. The dispute over what AMI spokesman Gerald McKelvey described as a public health hazard left a trail of bitterness and local concern.[144]

Blanco had spent twenty-three days in the hospital, and his recovery took four months. But he considered himself lucky, shuddering when he heard about the mysterious anthrax fatalities in New York and Connecticut and the series of misjudgments that left two of Washington's Brentwood workers dead. When he felt well enough to travel, Blanco and his family took a celebratory trip to New York

City, their luxury hotel paid for by the Bayer Corporation, the pharmaceutical giant that makes Cipro. Blanco eagerly endorsed the antibiotic, convinced that it had helped him survive the harrowing illness.

Blanco kept thinking that he would hear from the federal government. He didn't expect an apology; in the AMI case, none seemed necessary. He just believed that the government could learn from the survivors and that they deserved recognition. His daughter told him that Florida governor Bush had called to wish him well while he was in the hospital, but he could not remember it.

Finally, his doctor told him that a man was flying down from Washington to take him to dinner. The visitor was Dr. Arthur Friedlander, USAMRIID's senior anthrax researcher, who had led the seminal research on a controversial U.S.-produced anthrax vaccine.

In lab studies, Friedlander's unvaccinated lab animals had invariably died. But research on anthrax vaccines and medications had faced an insurmountable obstacle. Normally, once a drug's efficacy and safety is proved in animals, researchers test it in human clinical trials. With anthrax, there was no ready supply of human test subjects. Inhalation anthrax struck rarely, and it was too fatal even to contemplate infecting a human with it to test a vaccine's effectiveness.

As Friedlander explained, "There are not enough cases of inhalation anthrax in the world to test the vaccine, nor can we do challenge studies in human volunteers as we can do with other diseases, for which the mortality is lower and there's effective treatment. So we're sort of stuck—unless there's a cloud that comes over Washington and the people at the Pentagon survive and the others don't. It's very, very difficult with these vaccines to determine efficacy by the kind of standards that we would like to do and that we expect in other drugs, therapeutics, and vaccines, because of the rarity of the diseases."[145]

Now researchers had their first chance since the 1979 Sverdlovsk accident to investigate why some unprotected human beings could fight off the anthrax bacteria while others could not. They were eager to adjust their estimates about the quantity of spores needed to cause infection, obviously far fewer than the threshold that researchers had set before the outbreak, and to refine their evolving knowledge about the antibiotic regimen most likely to reverse the disease's course.

During the attacks, Cipro had proven to be more effective when used in combination with other potent antibiotics. The survivors offered a window into the unknown.

The effort to find patterns had begun almost immediately after the attacks. In December, 2001, a team of physicians involved in the care of anthrax patients, including Carlos Omenaca, the doctor who had treated Blanco, and the CDC's Zaki, Popovic, and Perkins, published a paper analyzing the first ten cases of inhalation anthrax linked to the attacks. The summary, which covered cases before Lundgren, made clear the importance of prompt antibiotic intervention.

In the clinical summary, the doctors noted a median patient age of fifty-six, with seventy percent of them male. Except for Kathy Nguyen, "all were known to or believed to have processed, handled or received letters containing *B anthracis* spores." Four patients had previously existing cardiovascular or cerebrovascular conditions; one had diabetes; one had asthma. None of them smoked.

"Eight of the ten victims were in the early stages of the disease when they sought care. Six received antibiotics the same day, and all of them survived," the report said.[146]

The study reached an encouraging conclusion that came far too late to save the two Brentwood workers. Previous estimates had suggested that a patient infected with inhalation anthrax had a fifteen percent chance of surviving. But with a cocktail of antibiotics begun during the early phase of the illness and "aggressive supportive care," the study said, those chances were "markedly improved." The doctors recommended further study of the survivors to better define the recommended treatment.

Despite the entreaties of prominent specialists, it took the federal government months to begin a systematic monitoring of the long-term effects of anthrax. Friedlander, one of the first to endorse the idea, made his first visits to Blanco and a few other survivors on his own, then became part of a fifteen-member team funded by the NIH to study the issue

Blanco, who hoped that the lessons learned from his struggle with anthrax would help future victims, didn't mind that he had become a kind of human guinea pig.

Neither did Leroy Richmond, although he had hoped for something more from his government, some tribute to what he and the others had endured. It stung when he watched Postmaster General Potter unveil a stamp commemorating the heroes of the fall's terrorism attacks, an effort to raise money for the families of rescue workers killed during 9/11.

Not a word was said about the men and women of the USPS who had survived a different kind of terrorism. No one seemed eager to take up their cause.

Soon after the anthrax scare subsided, a clamor arose for a vaccine that could protect postal and other high-risk workers from further anthrax attacks. Health officials reported that no vaccine was commercially available to inoculate U.S. citizens. The reasons were not medical or scientific—a Michigan operation licensed by the federal Food and Drug Administration had been making a vaccine used by the Pentagon for more than two decades. But in recent years, the manufacturer, BioPort Inc., had been sinking in a political and financial quagmire that threatened to put it out of business. The anthrax letters suddenly reversed the company's fortunes, making it one of the early beneficiaries of the nation's first major bioterror attack.

BioPort's saga began in 1997 when the Michigan Department of Public Health decided to unload its money-losing vaccine manufacturing plant, the Michigan Biological Products Institute, based in Lansing. The facility produced an anthrax vaccine used until the early 1990s by only a few thousand people, including veterinarians, millworkers, and researchers like USAMRIID's Friedlander and Ezzell.

The facility had run into problems with the FDA over quality-control issues, and expensive renovations had been ordered. The timing for a sale seemed right. The Department of Defense, alarmed by the biological threat that had surfaced in the Persian Gulf War, had decided to require anthrax vaccinations for all U.S. servicemen. Attracted by the prospect of lucrative DOD contracts, twenty companies bid on the Michigan facility. The state narrowed the field to two, finally settling on a $25 million bid from BioPort, a new venture, which boasted an impressive Pentagon connection. It was partly owned by William Crowe, a former chairman of the U.S. Joint

Chiefs. Crowe's involvement helped state officials overcome objections to possible national security risks posed by another major BioPort investor, Fuad el-Hibri, a foreign national of Lebanese descent with business ties to Saudi Arabia.

When BioPort won the bidding, Michigan State Representative Lingg Brewer filed suit charging that insiders from the state-owned lab had undervalued its vaccine contracts, then entered into a business partnership with BioPort. Brewer's lawsuit was dismissed, but he kept up a tireless campaign to discredit the company.

In 1998, BioPort sealed a $45 million vaccine contract with the DOD. The company had purchased vaccine stockpiles left behind by the state-owned lab, and these products were shipped out to begin vaccinating soldiers. BioPort hoped to produce fresh supplies by the end of 1999. First, it had to spend about $16 million on plant renovations to convince the FDA that it could meet its requirements. When the FDA came to inspect in November 1999, however, BioPort failed to win approval.

While BioPort regrouped, the Pentagon kept the money rolling in. The DOD gave BioPort an interest-free advance payment of $18.7 million and sweetened the deal by adding another $24 million to its contract. The price of a vaccine dose rose from an estimated $4.36 to $10.64. Critics charged that the Pentagon was propping up a favored contractor, unable to produce a safe product. BioPort President Robert Kramer said the revised contract was simply compensating BioPort for "our effort to get to FDA approval."[147]

Meanwhile, reports surfaced suggesting BioPort misused taxpayer money. A DOD audit questioned expenditures for furniture and office equipment and a chunk of money set aside for anticipated senior management bonuses, although Kramer said those payments never materialized. On Capitol Hill, Representative Walter Jones (R-N.C.) asked at a hearing, "How much longer are we going to permit the taxpayers of America to pay a company for a product it can't produce?"[148]

BioPort's problems escalated with mounting evidence of adverse vaccine reactions. In July 2000, BioPort employee Richard Dunn, sixty-one, died after getting a shot, and a medical examiner linked the death to the vaccine. Dunn's wife, Barbara, testified in October 2000 at a hearing of the House Government Reform Committee, which had emerged as a champion of soldiers claiming adverse vaccine effects. Mrs. Dunn reported that her husband had received eleven vaccine doses, the final two causing constant nausea, joint pain, and swelling. After her husband's death, she said that BioPort representatives "went on television and said they had no idea Dick had ever shown symptoms." She said his side effects were similar to those others had reported after receiving BioPort's vaccine.[149]

The committee heard complaints from a long line of soldiers who had been vaccinated, like army sergeant Robert Soska, who experienced muscle spasms and excruciating pain, and air-force captain Jon Richter, who had aching shoulders and spinal pain. Richter said he would resign before taking another required anthrax injection. In March 2001, the committee called for an end to mandatory vaccinations. Still, servicemen who refused faced court-martial.

The vaccine debate took on all the elements of Washington intrigue, dampening the chances that other pharmaceutical companies would take the risk of developing an alternative product and challenging BioPort's hold on the market. Complaints mounted among soldiers who had been vaccinated with a batch of BioPort vaccine that was later recalled by the FDA for incorrect expiration dates. A theory took hold that the anthrax vaccine could be at the root of the mysterious Gulf War syndrome, a collection of rashes and muscular ailments that generated more than fifteen hundred claims from veterans. Vaccine proponents like USAMRIID's Friedlander, who had taken shots without ill effects for twenty-three years, contended that BioPort's product, while imperfect, without question lessened the risk for soldiers marching into enemy territory.

Nevertheless, by August 2001, weeks before the anthrax letters surfaced, Secretary of Defense Donald Rumsfeld said that BioPort's contract was under review and might not be salvageable. Already, the DOD had cut back on its plan to vaccinate 2.4 million active and reserve troops. Unable to secure private financial backing, BioPort relied totally on the DOD for its income. Senator Tim Hutchinson (R-Ark.), calling the company "an unmitigated disaster," had proposed legislation giving BioPort until April 2002 to win FDA approval. The company, he said, was "costing the American taxpayer millions and millions of dollars and jeopardizing the safety of our troops."[150]

Suddenly the emergence of the anthrax letters brought a change of climate. The FDA, responding to the public clamor for protection, said through a spokesman that it would do anything possible to help BioPort clear up its manufacturing problems while still "ensuring the safety and effectiveness of the vaccine."[151] Two months later, the FDA gave BioPort its long-sought approval.

BioPort's phone lines buzzed with public requests, and federal officials suggested making the vaccine available to workers especially vulnerable to exposure. BioPort said it might charge as much as $90 per dose if the vaccine was marketed to the public, prompting persistent critic Brewer to quip, "That place will make millionaires out of a few people yet. Then again, they may strangle themselves on their own greed and incompetence."[152]

On October 19, 2001, a class-action lawsuit filed on behalf of more than two dozen soldiers alleged that the BioPort vaccine had caused them severe harm and even death. One plaintiff was the family of army specialist Sandra Larson, of Spokane, Washington, who had fallen into a coma and died after receiving a shot. BioPort denied the lawsuit's allegations.

In March 2002, the National Academy of Sciences Institute of Medicine affirmed the vaccine's safety and efficacy against all forms of anthrax, including inhalation disease. Reviewers said they found "no evidence that people faced an increased risk of experiencing life-threatening or permanently disabling adverse events" after receiving the vaccine.[153]

To BioPort's great relief, the Pentagon announced in June 2002 that it would resume mandatory vaccinations. Dr. William Winkenwerder, assistant secretary of defense for health affairs, said that the threat of an anthrax attack against the military was "still very real."[154]

He also said that roughly half of BioPort's production would be stockpiled, along with large reserves of antibiotics, for use by civilians in the event of a domestic emergency.

After nearing collapse, BioPort was looking toward expansion. Kramer said the company would work to "better support civilian needs.

"We were all under, and still are, a tremendous amount of pressure to get enough product out there as quickly as we can. We have been manufacturing at full capacity since December of 2001 and we will continue seven days a week to produce and make available as much vaccine, under our internal quality controls, as well as FDA requirements," he said.[155]

In mid-December 2001, HHS Secretary Thompson announced that his agency would make BioPort's vaccine available to ten thousand people who had been placed on antibiotics to counteract exposure to the anthrax letters. The news brought little jubilation. Because the vaccine was FDA-licensed for protection before anthrax exposure, Thompson's plan to use it after exposure required that participants sign consent forms saying they understood the risk.

Among the estimated three thousand postal workers exposed to anthrax, the offer precipitated a storm of racially charged criticism. William Smith, the president of the New York Metro Area Postal Union, dashed off a letter to Thompson comparing the offer to the notorious Tuskegee syphilis study that the U.S. Public Health Service performed between 1932 and 1972 in Macon County, Georgia, in which hundreds of African-American men suffering from syphilis were denied treatment while researchers observed their decline.

A *Washington Post* editorial pointed out that "to ask people already struggling with the emotional fallout of anthrax exposure to make the complicated medical decision to join a clinical study is to impose a tremendous burden." Postal workers, the editorial stated, were most likely to face this burden. To enter an experiment as a human subject is not a decision to be taken lightly, even with a vaccine that, like this one, has shown no serious safety problems. Access to the vaccine option must not translate into pressure to take it. Otherwise, those on what turned out to be anthrax's front line are being twice victimized."[156]

Thompson's offer ended with a thud. Of the ten thousand people eligible, only a hundred and thirty chose to take the vaccine, forty-eight of them congressional employees. To the puzzlement of federal health officials, only six Brentwood postal workers agreed to take BioPort's controversial product.

In the aftermath of the anthrax attacks, the Hart Senate Office Building was the object of the first phase of cleanup, a $23 million job that involved spraying the grand atrium and beautiful office suites with a noxious solution of chlorine dioxide gas. Environmental companies traveled to Washington from distant points for slices of the cleanup pork. Few questioned the cost, although a later review showed that totals had skyrocketed as work was contracted without competitive bids.

Staffers moved back into the building on January 22, 2002, with EPA assurances of safety. But after dozens of reported illnesses, authorities had trouble sorting out whether the cause was the lingering smell of the cleanup, the overly irradiated mail, or simply the post-traumatic stress of returning to the building.

The Brentwood cleanup offered a much larger challenge. At the Hart Building, only 100,000 cubic feet had to be sealed off for fumigation. Brentwood, on the other hand, had 17.5 million cubic feet to clean up. Experts considered it an unprecedented reclamation of a biohazardous building. The project called for extensive planning by USPS authorities, who estimated that cleanups at Brentwood and Hamilton would cost $35 million.

The first part of the project was dealing with the abandoned mail, 2 to 3 million pieces at Brentwood and another million or so at Hamilton, which was considered dangerous. Federal officials had learned their lessons about possible cross-contamination from the strange deaths of Nguyen and Lundgren, so the USPS began shipping truckloads of mail in sealed containers to the Titan Corporation decontamination facility in Lima, Ohio.

The first phase, in which letters were sterilized with an electronic beam, was completed by February 2002. However, decontaminating the flats, or oversized mail pieces, lasted until mid-March.

Packages required even more time. Concerned that the beam would not penetrate deeply enough, the USPS waited until early June to ready a new facility in Bridgeport, New Jersey, that could use an X ray to sterilize the packages. Throughout the months after the attacks, the USPS also irradiated all new mail addressed to government office buildings in Washington, adjusting the dosage downward after 255 illness complaints from Capitol Hill.

But there were challenges beyond decontaminating the backlogged mail. Just getting inside Brentwood was a complex procedure, like entering a BSL3 laboratory at USAMRIID. In charge of cleanup efforts was Thomas G. Day, the USPS vice president for engineering. Day contracted the job out to biohazard companies. "You don't just put on a protective suit and walk through a door into the building. Obviously, if you just opened the door to the building, it would allow anthrax spores to get out," he said.[157]

The spectacle around Brentwood heightened neighborhood anxiety about the potential dangers of the next phase: the infusion of chlorine dioxide gas to kill latent anthrax spores. Residents and some health authorities worried that in a strong wind, chemicals would spread over the area and sicken families.

Many Brentwood workers who had been reassigned to other postal centers wanted no part of a cleaned-up building. Some postal union officials expressed the wish that the USPS would tear the place down.

Instead, the USPS vowed to bring Brentwood back to life, cleaner than ever and fully renovated. On July 29, 2002, the agency did a limited test inside a tent it had constructed within the plant. Carbon dioxide was held inside the plant for twelve hours at a specific temperature, humidity, and concentration. Samples were taken from the most contaminated parts of the building, Day said, and when the results came back "we had a hundred percent kill rate."[158]

After attending a series of community meetings and gathering approvals from government agencies, Day planned to begin pumping the gas into Brentwood. When this phase was completed, he would dismantle the whole operation and transport it to Hamilton. He hoped to reopen Brentwood early in 2003.

"We'll be able to open up the building again," Day said confidently. "And actually, before we let our people go back in there to work, we're going to refurbish it a bit. That'll probably take an extra couple of months to get it all cleaned up, recarpeted—just basic general housekeeping things—so that we can give our employees the kind of workplace they deserve." Day said that he and other postal officials, while determined to reopen the plant, were sensitive to workers' fears about returning. Along with D.C. Representative Eleanor Holmes Norton, Day said he would be among the first to reenter Brentwood as a symbolic show of confidence.

Day said he wanted to set an example, showing "a level of compassion and understanding to the anxiety" but confidence in the science.

Postal officials planned one other symbolic concession to Brentwood anxieties. Although it had been decontaminated in the July test run and offered some recyclable parts, Machine No. 17—the mail sorter that had infected Curseen and Richmond—would be dismantled and removed before Brentwood reopened.

"I think you understand," Day said. "That was the machine [the Daschle letter] was processed on. And there is anxiety about it. I think it is the right thing to do for our employees to put a new one in its place."

CHAPTER 15

A Person of Interest

August 25, 2002
ALEXANDRIA, VIRGINIA

"I want to look my fellow Americans directly in the eye and tell you that I am not the anthrax killer. I know nothing about the anthrax attacks. I am an innocent man."

Dr. Steven Jay Hatfill scowled into the midday sun outside his Virginia attorney's office, his gaze fixed on a ring of television cameras set up along the quaint brick sidewalks. For his second press conference in a brutally hot stretch of August, Hatfill had attracted a pack of sweating, irritable reporters from all over the world to witness a daring legal defense strategy, unprecedented in its show of raw egotism and righteous indignation. He had launched an aggressive preemptive strike against the FBI, declaring his innocence before he had been arrested or even confirmed as a suspect in five anthrax deaths. Square-jawed and sturdily built, with military bearing and a tiny American flag affixed to his lapel, Hatfill called for an end to a campaign of FBI harassment and leaks to the news media that had ruined his career, a flamboyant journey through the fields of science, soldiering, medicine, and bioterrorism consulting. Hatfill described himself as an American patriot, a healer devoted to saving lives, not taking them.

Hatfill's effort to portray himself as a victim of an FBI vendetta had the staged feeling of a Hollywood movie shoot. Pretty female legal aides passed out thick press packets, each containing CD-ROM photographs of his girlfriend's apartment, which had allegedly been ransacked during an FBI search. Pat Clawson, a media-savvy former CNN reporter who had appointed himself an adviser to Hatfill's defense team, led the somber doctor on a slow stroll toward historic King Street so cameramen could get the perfect live-action shot. Dour but handsome at forty-eight, with wide shoulders and blue eyes, Hatfill detoured into the tourist-lined street to shake hands with amused Alexandria police officers.

Hatfill's rehearsed twenty-minute diatribe against the FBI accused the Bureau of hounding him, and the U.S. Attorney General's Office of a "never-ending torrent of leaks and innuendo."[159] He lashed out at *New York Times* columnist Nicholas Kristoff, who had criticized the FBI for failing to arrest an anonymous "Mr. Z," whose description read like Hatfill's biography.[160]

Hatfill also attacked Attorney General Ashcroft, recalling Ashcroft's own struggles to win Senate confirmation. This seemed especially ill-advised since Hatfill, in July 2002, had managed to land a cushy $150,000-a-year bioterror training job at Louisiana State University funded through a Justice Department grant. Even though the FBI had trailed Hatfill to Baton Rouge, LSU's continuing-education division had risked hiring Hatfill because he had not been designated a suspect in the investigation, a usual precursor to indictment. Ashcroft referred to Hatfill publicly only as a person of interest in the anthrax probe, one of several whom he said the FBI had had under surveillance for months.[161] The phrase *person of interest* carried no legal meaning, and Hatfill sneered as he repeatedly used it to describe himself during the sidewalk screed.

As a "person of interest," he said, his life had been shattered, his scientific accomplishments besmirched, his new LSU job jeopardized.

> *A person of interest is someone who comes into being when the government is under intense political pressure to solve a crime but can't do so, either because the crime is difficult to solve or because the authorities are proceeding in a wrongheaded manner. It then becomes necessary for the*

> *authorities to produce a warm body. But since there is no suspect and the authorities have nothing on which to base a prosecution, they pick a serviceable target. This should preferably be a person about whom mysterious questions have been raised. They give him a prejudicial label—"person of interest"—and leak inappropriate rumor and innuendo to the press. They then sit back and watch an uncharged, presumptively innocent person be picked apart by journalists looking for hot stories.*

Hatfill said that agents had told his girlfriend he was responsible for the deaths of five innocent people. To disprove this, he provided reporters at the press event with time sheets showing work hours that he had logged on the days the Daschle and Leahy letters were mailed. He offered to give handwriting samples and take a blood test to show that his body had produced no anthrax-fighting antibodies.

Finally, Hatfill stepped back from the cameras and embraced his spokesman and friend Clawson in a clumsy bear hug. The men slapped each other's back awkwardly. Hatfill grimaced as the cameras panned his face, his eyes closed dramatically as if to fight back tears.

Although he spoke defiantly, Hatfill had become convinced that he faced arrest, perhaps on a charge that would allow authorities to hold him in protective custody. Only days before the media event, agents had flashed Hatfill's picture at shopkeepers in Princeton, New Jersey, asking if anyone had seen him. They had pressured his Malaysian-born girlfriend to become a government witness, also leaning on her relatives in New Jersey.

It should have been easy enough to find a way to detain or pressure Hatfill. His résumé suggested he had made sworn statements to government agencies that were provably false, a crime under federal law, and managed to win the "secret" clearance needed to enter USAMRIID and other government labs and military installations. But the FBI faced embarrassment in focusing on such breaches unless agents could show real evidence related to the case.

Although Hatfill had passed himself off as a member of the U.S. Army Special Forces, the Pentagon had dredged up records showing he had failed the training and been forced to drop out. While working at federal research labs, Hatfill claimed to have a Ph.D. in molecular biology, although he had not yet earned the degree. Officials later said he had submitted a certificate, later discovered to be fabricated, showing he was awarded the doctorate. It took two years for federal authorities to revoke the security clearance he had won using these false credentials.

Hatfill professed to be an innocent man, ruined by a campaign of government harassment. Yet, he said, even he could see why the FBI might single him out. In many respects, he fit the psychological profile drawn up by FBI behavioral experts of the type of person likely to be involved in the anthrax crimes.

Still, Hatfill's attack on the FBI raised a salient point. It was true that the FBI had been under pressure to solve the case, both from Congress and the general public. Already under fire for its failure to detect and thwart the plans of the 9/11 hijackers, the bureau badly wanted to prove its competence by making an anthrax arrest.

The case, however, had presented unusual obstacles. The scientific and technical aspects of the attacks had pushed the bureau to the limits of its abilities. For help, it had relied heavily on outside experts like John Ezzell and his USAMRIID colleagues. Yet these same experts were among an elite cadre of scientists with the requisite knowledge to carry out the crimes. The overlaps between the list of experts and the list of possible suspects were alarming.

The attacks—with letters mailed from different points on the map and attention lavished on the smallest detail that could throw investigators off the trail—seemed to be the work of more than one person. The notion that the hoax letters might also be the work of the same perpetrator spread the geography even farther—since letters had been postmarked in Malaysia, Britain, and elsewhere. Yet if more than one person had been involved, risking death or illness, then it was even more mystifying that solid evidence had not turned up.

Anyone involved, experts said, would have had to take medical precautions—dosing themselves with either the anthrax vaccine or antibiotics. On this assumption, the FBI looked for prescription records. The perpetrator or perpetrators, experts said, would have needed access to specialized equipment sold to a discrete group of scientific customers. The FBI reached out to equipment suppliers and questioned USAMRIID officials about three discarded biosafety cabinets Hatfill had been allowed to take from the lab. The culprit must have worked in a remote location, and the site likely would have become so contaminated that he would have had to burn it to the ground to destroy potential evidence. So the FBI perused fire and arson reports.

Some attempts to shoehorn Hatfill into a matrix failed. Agents thought they had a promising lead when they heard that Hatfill had frequented a remote cabin in the Shenandoah Mountains the summer before the attacks and, at one point, had discussed putting friends on Cipro. The FBI grilled the Washington lawyer who owned the "cabin," which turned out to be a well-appointed house, and others who had used the property.

Hatfill had discussed Cipro, but in a different context, according to his spokesman. During one of their male-bonding sessions at the mountain house, Clawson had told Hatfill about a letter containing harmless powder that had been sent to Iran-Contra figure Oliver North at Radio Liberty. This led Dr. Hatfill to discuss the treatment of anthrax exposure and other possible uses of Cipro.

Clawson said the freewheeling chat led inevitably to the subject of sex. It was decided that they should all go on Cipro since it might be an effective remedy for treating "the clap."[162]

Hatfill had a flair for the dramatic. Once, he had draped himself in plastic garbage bags and a face mask to show how easy it would be to use simple items to produce biological weapons in the average kitchen. He spent years conducting medical and scientific research that made its way into prestigious journals; then he suddenly positioned himself as a bioterror expert who would lead American law enforcement to a heightened state of preparedness.

Hatfill's résumé impressed casual readers with the unorthodox twists and turns of what he once described as a "peripatetic career."[163] He joined the Rhodesian army when the southern African nation, now Zimbabwe, was under apartheid rule. He boasted that he served in its notorious Selous Scouts, a mercenary force in operation from 1973 to 1980 that was later implicated in biological warfare experiments and other atrocities.

He had explored Antarctica from a South African research station built atop the polar ice. He won certification in underwater medicine and advised NASA on its manned space flights. Years later, returning to the United States after sixteen years abroad, he helped investigate the deadly Ebola and Marburg viruses in USAMRIID's BSL4 laboratories.

The falsified résumés that he circulated to federal agencies offered a glimpse of the daring man whom colleagues saw as brilliant yet difficult, with a huge ego, an abrasive personality, and a bad temper. They also showed why the FBI would single him out for scrutiny.

Amid the inflations of a 1999 résumé, Hatfill cited his familiarity with "former U.S. and foreign BW [biological weapons] programs, wet and dry BW agents, large-scale production of bacterial, rickettsial and viral BW pathogens and toxins, stabilizers and other additives, former BG *[Bacillus globigii]* simulant production methods, open air testing and vulnerability trials, single and 2 fluid nozzle dissemination, bomblet design, munitions programs, and former Soviet BW programs."[164] It was hard to know whether Hatfill knew anything more about bioweapons than what he might have picked up while hanging out with bioweapons expert Bill Patrick, who was listed as one of his references, or Russian defector Ken Alibek, whom he knew through the bioterror lecture circuit.

Hatfill was born in St. Louis, Missouri, in October 1953, the son of a horse trainer who eventually moved to Florida and raised Thoroughbreds in Ocala. In recent years, Hatfill sometimes listed the family farm, Mekamy Oaks, as his own home address, and he kept belongings in a rented storage shed there. Intrigued by the Florida connections to the anthrax death of Bob Stevens and to two hoax letters mailed from St. Petersburg in October 2001, FBI agents during the summer of 2002 sifted several times through Hatfill's Florida shed. They carted away bags of materials, although agents would not disclose their contents.

In his youth, Hatfill studied biology at the small Southwestern University in Winfield, Kansas. He took a break from college to work in Zaire at a medical clinic run by Dr. Glenn Eschtruth, a Methodist missionary. Hatfill married Eschtruth's daughter and talked about returning to the remote clinic. Those plans fell apart when Eschtruth became the only American killed by mercenaries in that country's civil conflict, prompting speculation, never confirmed, that Eschtruth had been a CIA operative.

Hatfill divorced, and instead of pursuing a medical degree after college, joined the military. He wanted to serve in the army's Special Forces, but he flunked out of the training at Fort Bragg. If his résumés are to be believed, he served simultaneously in the U.S. Army Institute for Military Assistance in Fort Bragg and the Rhodesian Special Air Squadron. Records in Zimbabwe confirmed he served in the Rhodesian military, but his assertion that he served in the Selous Scouts remained in dispute.[165]

Although Hatfill was in Rhodesia in 1979 at the time of a widespread anthrax outbreak in the Tribal Trust Lands that killed 182 people, no evidence surfaced showing that he had any involvement in what was later judged to be a stealth anthrax attack in the African bush. He later chastised reporters for printing innuendo based on happenstance. Similarly, he acknowledged having once lived in a suburb of Harare, Zimbabwe, known as Greendale, but insisted that the news media jumped to conclusions by suggesting that the address could be connected to the nonexistent Greendale school used in the return addresses of the Daschle and Leahy letters.

Hatfill graduated from the Godfrey Huggins School of Medicine in Rhodesia in 1984, then ventured to Antarctica to lead a South African research team on the Fimbul Coastal Ice Shelf of Queen Maud Land. He shared his studies of the physical effects of the Antarctic midwinter syndrome with NASA; he had a lifelong dream to one day explore Mars.

He earned two postgraduate degrees at the University of Stellenbosch Medical School in South Africa and worked in its radiology lab. With an article in the *Lancet,* and invitations to speak to various professional conferences, Hatfill's medical credentials grew. He submitted a dissertation for a Ph.D. in molecular cell biology at Rhodes University in South Africa but it did not pass the committee review.

Hatfill studied for a year at Oxford University in England, then returned to the United States. In 1995, he won a two-year NIH fellowship, becoming part of a team analyzing the links between the human spleen and T cells.

When the fellowship expired, Hatfill won a similar position with the National Research Council (NRC), which dispatched him to study viruses for two years at USAMRIID, a job requiring security clearance. Despite its sensitive mission, USAMRIID relied on the NRC to screen fellows for risks.

David Franz, the former USAMRIID commander who had served as a Green Beret, said he remembered once chatting with Hatfill about his Rhodesian military service when they met in USAMRIID's hallways. "I had no reason in those hallway discussions to even think that he might not be telling the truth," Franz said.[166]

In September 1997, Hatfill completed USAMRIID training for BSL4 lab work and also received the anthrax vaccine. He took special training to work with USAMRIID's Aeromedical Isolation Chamber, which allows medical teams to evacuate bioterror victims without touching them. Hatfill later described himself as the "primary civilian consultant" to the team.[167]

The USAMRIID course marked the start of a career shift. Hatfill came out of the laboratory into the limelight. Bioterrorism had suddenly become a hot topic, with opportunities for both publicity and profit. Some scientists viewed the field of "bioterror experts" with disdain.

"There are people who I call 'Chicken Littles' who go around saying, 'You know, the sky is falling! The sky is falling. We're all going to die. Someone's going to spray some powder or smallpox or whatever, and everybody's going to die!'" grumbled Abigail Salyers, the former American Society for Microbiology president.[168] Whether they were opportunists or prophets, bioterror "talking heads" were in even greater demand in the late 1990s as the Pentagon fretted about the weapons capabilities of Iraq and as hoax letters escalated public concern.

Hatfill seized a piece of the action. In a *Washington Times* column published in August 1997, he described four levels of bioterror threats and discussed the ease of producing a widespread panic. He cited as an example the still-unsolved 1997 anthrax hoax targeting the B'nai B'rith headquarters in downtown Washington that caused the evacuation of hundreds of people.

By 1998, Hatfill had appeared on the television talk show hosted by conservative Armstrong Williams, stressing his conviction that a bioterror attack was imminent.[169] A magazine featured his lesson on preparing bioterror agents in the kitchen. He gave briefings to the FBI's Joint Terrorism Task Force, the CIA Counter-Terrorism Center, military officials, and defense agencies.

Some of his appearances teamed him with Patrick and Alibek, the most respected names in the field. In June 1998, Hatfill was listed with them at a seminar hosted by George Washington University's Terrorism Studies Program and the Potomac Institute for Policy Studies. Hatfill described the military's need to develop equipment like a "disaster evacuation train" that could be used to transport massive casualties. He also showed slides of the types of infections that could be expected after an anthrax attack.[170]

When he joined Science Application International Corporation (SAIC) in early 1999, Hatfill maintained close contact with Fort Detrick. At one point, he returned to USAMRIID to pick up the outdated safety cabinets, arousing interest when he said he would use the equipment for a secret training exercise.

In the year before the anthrax attacks, Hatfill's career wavered between new highs and lows. He alienated some colleagues by questioning the credentials of an in-house rival at SAIC, who responded by scrutinizing Hatfill's credentials. He applied for a CIA job requiring an enhanced security clearance. His unsatisfactory answers to CIA inquiries cost him this clearance in late 2001. SAIC removed him from his full-time job but kept him on as a consultant.

Through his own initiative, however, Hatfill managed to earn a place as a trainee for a United Nations inspection team that would be ready to move into Iraq at a moment's notice to verify its weapons capabilities. He had offered his services to the UN Monitoring, Verification and Inspection Commission (UNMOVIC) as an experienced hand in working with biological agents. The UN Security Council in December 1999 established UNMOVIC, as a successor to UNSCOM, to implement and monitor the disarmament of Iraq. Most nominees came in through official channels. Hatfill was among the few to forward his own name. He submitted a standard personal history form and résumé. An UNMOVIC spokesman said the commission accepted these documents at face value, lacking the resources to verify the credentials.[171]

The most stringent requirement was a lengthy interview at the UN headquarters in New York, usually conducted by UNMOVIC's executive chairman, Dr. Hans Blix of Sweden. Hatfill was accepted. In April 2001, he joined a dozen trainees for a session in biological weapons detection at the Edgewood Arsenal, in Aberdeen, Maryland, an army test site with a containment chamber in which live agents could be used to measure the effectiveness of biowarfare equipment.

In mid-November 2001, Hatfill traveled at UN expense to Swindon, England, a suburb about an hour and a half outside London, for more intensive training. The UN course ran from November 12 to 23, and the trip would later take on significance as the FBI retraced Hatfill's steps. The training included field trips to facilities using specialized equipment that might be seen in an Iraqi bioweapons production plant.

Hatfill returned to the United States as the anthrax scare climaxed with the discovery of the Leahy letter and the Lundgren death. Hoax letters continued to surface around the world. In early January, a hoax anthrax letter was found in Zimbabwe at a Harare post office. The letter contained a material that sickened two postal workers, followed by similar letters at two other sites in Southern Africa.

Not long after John Ezzell analyzed the Daschle letter in mid-October, federal law enforcement officials quietly acknowledged that the perpetrator of the first domestic bioterrorism attack most likely was not a foreigner but an "insider" with connections to a U.S. defense research facility. Homeland Security Director Ridge publicly confirmed this on January 13, 2002, when he said that the focus of the investigation had "turned inward."[172]

The United States hoped that the Iraqis had not acquired the Ames strain, although it was impossible to tell whether samples had somehow ended up in enemy hands. The flow of detailed information about Iraq's bioweapons capabilities had ended abruptly when Saddam Hussein ousted UN inspectors, and concerns had mounted in the summer of 2001 after a CIA report suggested that Iraq might again be developing biological weapons. In September 2001, Defense Secretary Rumsfeld had told a television interviewer that he believed the Iraqis had been "working diligently to increase their capabilities in every aspect of weapons-of-mass-destruction and ballistic missile technology."[173]

Few government scientists had access to the Ames strain, but even fewer had worked with anthrax aerosols. USAMRIID anthrax researchers used only liquid anthrax, not powder, and opinions varied about how difficult it would be to steal leftover spores from a USAMRIID nebulizer, culture them, and convert them off-site into a weapons-grade powder.

A breakthrough seemed imminent when officials at the army's Dugway weapons test center acknowledged that for years its scientists had created small quantities of aerosolized anthrax to assess battlefield detection devices and protective equipment. This opened the possibility that the anthrax used in the letters had simply been stolen from a military site. Dugway's creation of dry anthrax started after the Gulf War, when scientists stepped up their efforts to counter the perceived Iraqi threat. The powdered anthrax had been made under Bill Patrick's tutelage, using a freeze-dried technique.

After concerns subsided that Dugway was producing material banned under the 1972 weapons control treaty, little more was said about the dry anthrax. Production quantities were not divulged, nor was the list of outside parties who had received the powder. The FBI moved into Dugway, interviewing several hundred employees and subjecting some to polygraph tests. Dugway officials declined to answer questions about whether powder samples were ever shared with USAMRIID or with the army's testing facilities in Aberdeen.

Meanwhile, analyses of the Daschle and Leahy anthrax offered insight into the techniques that might have been used by the killer to create a powder. A report to the FBI from one of its outside laboratories suggested that the Leahy anthrax had been processed with a spray dryer, an expensive, specialized machine commonly used by pharmaceutical companies. Spray drying yielded a consistent particle size of one to three microns, which was fine enough to be sniffed up the nose and penetrate the lungs.

One analysis suggested that particles of the Daschle/Leahy anthrax had been coated with silica to reduce its electrostatic charge. The silica also appeared to contain a chemical additive to aid in bonding, not unusual in pharmaceutical manufacturing. Other scientists looking at the material came to different conclusions.

The FBI's effort to keep track of viable suspects was just as challenging as the technical issues. In late September 2001, the FBI had received an anonymous letter suggesting that a former scientist at Fort Detrick could be a terrorist planning some sort of attack. The letter named the Egyptian-born Dr. Ayaad Assaad, who had worked for the army's Medical Research Institute of Chemical Defense on research into the plant toxin ricin, a potential biological warfare agent. After Assaad was forced out of Fort Detrick by a staff cutback, he and several other USAMRIID scientists had filed a lawsuit accusing their employer of harassment.

The FBI eventually interviewed and cleared Assaad. When details of the accusations surfaced in the press, he and his codefendants won an audience for their complaints about Fort Detrick working conditions and security lapses. In the lawsuit, Assaad accused a colleague, a microbiologist and rival ricin researcher, of waging a campaign of ethnic slurs and organizing a "Camel Club" to poke fun at Assaad's Egyptian heritage. The club's mascot was an obscene rubber camel, used as an award for the week's worst performer.

Assaad's tormenter had left Fort Detrick, but Assaad and the codefendants contended in press interviews and in the lawsuit that the scientist continued to make unauthorized visits to Fort Detrick's labs in the early 1990s, sometimes late at night, in the company of a female USAMRIID colleague.

David Franz, who was deputy commander during the early 1990s, disputed allegations of faulty security. "You had to have a card to swipe to get into a suite. You had to have a memorized PIN number," he said. "You had to have passed muster with your division chief. . . . If your card went in, and your PIN number was typed in, we knew it—every time. We had records of everyone who went into there, twenty-four hours a day, how long they stayed there, and so on and so on."

The year 2001 ended with no arrests and no obvious suspects. But one new intriguing piece of evidence did surface in January: a second letter that had been mailed to Senator Daschle's office in the Hart Building. Although it was postmarked in late November and proved to be a hoax, its delivery to the Daschle office suite had been delayed by the irradiation of mail on Capitol Hill.

The letter resurrected fears inside the Hart Building, which had been reopened after three months of cleanup. More interesting than its contents was its postmark: it had been mailed from London, the only letter to originate in Britain. No other details were released.

In drawing up their initial lists of scientists to interview, FBI agents had been given Hatfill's name by several sources. They moved forward slowly and finally connected with Hatfill in early 2002. One reason he rose to the top of the FBI's list was the coincidence of timing: Hatfill had been in England in late November for UNMOVIC training. Hatfill's friend Clawson said the doctor had no knowledge of the hoax letter mailed from London around that time. He said Hatfill flew into Heathrow Airport, then drove directly to his suburban hotel and back again, with no detours into London.

In his initial interview with the FBI, Hatfill said he answered all questions and agreed to let agents look around his Frederick, Maryland, apartment without a search warrant.

Publicly, the FBI seemed to be grasping for leads. It blanketed the Trenton, New Jersey, area with flyers asking citizens to report any suspicious sightings. It also called on microbiologists for help. The FBI wanted a list of ASM's members, which number more than thirty thousand in the United States, so that it could begin contacting them for possible leads. This touched off concern within the ASM board of directors over the civil liberties issues raised by turning over privately held membership information to criminal investigators.

The FBI's Harp called Abigail Salyers to ask for help. He described how the FBI profile had narrowed the field, suggesting that the killer was "a loner and a loser, and probably a nerd," Salyers recalled.[174]

"That describes at least half of our members!" she explained to Harp. "Searching for someone like that in the American Society for Microbiology is like looking for a tree in a forest. The profile was so vague that I didn't think it was going to be much use."

Salyers urged Harp to send a letter to ASM members asking them try to "dredge up some memory that might have gotten obscured in the initial shock response."

Harp sent out the message on January 29, 2002, asking ASM members to forward leads to the FBI. "It is very likely that one or more of you know this individual," he wrote. "It is important to ensure that all relevant information, no matter how insignificant it may seem, is brought to the attention of the investigators in this case."[175]

ASM lost a couple of members over its decision to turn over the list, but the overwhelming majority supported the board's decision to cooperate. Scientists did what Harp had requested, searching their memories and notebooks for any encounter or contact that might have raised suspicion. But some regarded the FBI's appeal as a desperation tactic.

When the ASM gathered in May 2002 for its annual convention at the Salt Palace Convention Center in Utah, the FBI's early interest in the group seemed to have totally abated. Although the challenges raised by the anthrax attacks were a major focus of the meetings, there was no evidence of any FBI presence on the convention floor. Salyers and her successor, newly elected ASM president Ronald Atlas, said the FBI had not asked to be part of the conference.

USA Today Justice Department reporter Toni Locy had come to the forum expecting to find FBI agents scouting the crowd. When she did not, she set out on a short walk to the FBI's Salt Lake City field office to ask about it. She found agents comfortably settled in after the day's big event, the announcement of a drug-conspiracy indictment. They seemed surprised when Locy said there was a huge convention of microbiologists discussing anthrax only a few blocks away.

One scientist, Barbara Hatch Rosenberg, had already emerged as a major FBI irritant. She taught environmental science at the State University of New York at Purchase, and in 1989, she started a chemical and biological weapons program within the Federation of American Scientists (FAS) to work toward strengthening the 1972 biological weapons treaty.

Rosenberg rigorously tracked every evolving clue related to the anthrax letters and wrote analyses posted on the FAS Web site. She included hoaxes in her study, delving for any suggestion that they could have come from the same perpetrator.

Her first analysis contended:

> *The FBI has surely known for several months that the anthrax attack was an inside job. A government estimate for the number of scientists involved in the U.S. anthrax program over the last five years is 200 people. According to a former defense scientist, the number of defense scientists with hands-on anthrax experience and the necessary access is smaller, under 50. The FBI has received short lists of specific suspects with credible motives from a number of knowledgeable inside sources, and has found or been given clues . . . that could lead to incriminating evidence. By now the FBI must have a good idea of who the perpetrator is. There may be two factors accounting for the lack of public acknowledgement and the paucity of information being released: a fear that embarrassing details might become public, and a need for secrecy in order to acquire sufficient hard evidence to convict the perpetrator.*[176]

She went on to discuss the Ames strain and its distribution path, her belief that no more than four U.S. labs had the ability to weaponize anthrax, and her theory that the Ames sample most likely came from USAMRIID, Dugway, or Battelle.

Rosenberg also took the FBI to task.

> *The perpetrator is surely too smart to believe that either the FBI's ludicrous recent actions or the White House protestations of ignorance mean that the authorities are not on to him. Blanketing Central New Jersey with fliers showing handwriting that was obviously disguised can't possibly evoke useful information, nor can letters to 32,000 American microbiologists, 31,800 of whom live in a different world from the perpetrator. This is no way to instill public confidence in the competence of the FBI.*[177]

At a February talk in Princeton, New Jersey, Rosenberg suggested that the FBI had narrowed in on a suspect but had deliberately delayed bringing him into custody, fearful of publicly identifying the suspect. She said the evidence pointed to a person who had experience handling anthrax and was protected by vaccinations and boosters. She also suggested that the culprit had access to classified information about the process that allowed spores to stay airborne.

"We can draw a likely portrait of the perpetrator as a former Fort Detrick scientist who is now working for a contractor in the Washington, D.C., area," Rosenberg said. "He had reason for travel to Florida, New Jersey and the United Kingdom. . . . There is also the likelihood the perpetrator made the anthrax himself. He grew it, probably on a solid medium and weaponized it at a private location where he had accumulated the equipment and the material.

"We know that the FBI is looking at this person, and it's likely that he participated in the past in secret activities that the government would not like to see disclosed," Rosenberg said. "I know that there are insiders, working for the government, who know this person and who are worried that it could happen that some kind of quiet deal is made that he just disappears from view."

Rosenberg's remarks made her persona non grata at the FBI Washington field office. At one point, Harp called to warn her that her comments could damage the investigation. At the White House, Bush press secretary Fleischer said Rosenberg seemed to be overreaching in suggesting there was a single suspect.

But he added, "The president would like to get this, obviously, resolved as a quickly as possible. The pace of justice is a methodical one. It's very important for them to build a case that will stand up in court that is thorough and is conclusive."

Rosenberg only pressed more aggressively. She updated her analysis on June 13, asserting, "nearly everyone who has followed the situation closely... knows who a likely perpetrator is. The FBI continues to claim that it has no suspects and few clues, but it continues to focus on biodefense scientists with anthrax experience."

She provided no names, but her analysis included specific criticisms and questions about the FBI's performance. She asked whether the FBI had pursued "ancillary evidence" about the Daschle hoax letter or connected prior hoaxes to the pattern.

Her persistence finally paid off. Rosenberg was invited to brief the staff of the Senate Judiciary Committee.

Harp and a contingent from his office also attended, and did not hide their disdain for her theories.

Rosenberg went through her analysis of the likely culprit in the anthrax attacks, a description that in many respects mirrored the FBI's behavioral profile. She described a possible connection to UNMOVIC. She refused to identify her suspect, even when Harp pressed her.

"Do you know who did this? Do you know?" one participant remembered him asking several times.

Rosenberg said she did not.

When the session was over, a former prosecutor in the room advised Harp that perhaps Rosenberg, who had spent months gathering tips, possible leads, and hunches from scientists, merited gentler treatment.

Harp contacted Rosenberg the next day. As she described it, agents continued to press her for a name, even though "both Harp and the other agents knew very well from the beginning who and what I had in mind. Trying to make me verbalize it was a form of intimidation."[178]

A few days later, with television cameras rolling outside, FBI agents made their first publicized search of Hatfill's apartment.

Around and around the investigation went over the next few months. Speculation would reemerge that the anthrax could be Iraqi in origin, and then it would swing back to the likely U.S. sources. By October 2002, the FBI said it had interviewed 4,700 people in the course of the investigation.

Hatfill, after his fit of bluster during August, grew silent. Fuming, he would call his buddy Clawson on a car phone from congested Interstate 270, convinced that he was surrounded on all sides by unmarked FBI cars. Hatfill lost his sense of humor, particularly with prison jokes, Clawson said. He flew back to Baton Rouge to clean out the apartment he had briefly rented until LSU fired him. He planned to move in with his girlfriend, still loyal to him despite the FBI's pressure.

As late as December 2002, the FBI was still on his trail. Tipped that Hatfill had once discussed how he would dispose of anthrax processing equipment, agents dredged a pond in a Frederick, Maryland, public park.

Hatfill waited and pondered his future. He had told reporters angrily that his career was ruined, but clearly, he was a man with many unexplored talents not listed on his long, partially fabricated résumé.

When the FBI, carrying a search warrant, raided his apartment a second time, agents confiscated the hard drive of Hatfill's computer and resurrected an unpublished 197-page bioterror novel, titled *Emergence,* that was also on file at the U.S. Copyright Office.

The novel deals with the outbreak of a mysterious and deadly disease that strikes the White House, sickening both the president and a U.S. senator, and spreads across the United States. Scientist Steve Roberts is sent to investigate an apparent act of bioterrorism, a journey that takes him to Antarctica, through the federal bureaucracy, and the labs of USAMRIID.

He finally tracks down the Palestinian culprit, who, with Iraqi backing, worked out of a home laboratory to produce *Yersinia pestis,* the germ that causes bubonic plague. The terrorist used fake credentials to acquire scientific equipment and spent another $387 on household commodities such as Q-tips, bouillon cubes, and a pressure cooker.

Rob Thomason, a *Washington Post* researcher who reviewed Hatfill's manuscript, described it as filled with disdain "for all types of people except well-trained scientists." Hatfill portrays doctors at the CDC as a collection of social workers gathering irrelevant statistics. Journalists are sensationalistic, but more adept at getting information than government intelligence agents. The novelist contends that members of Congress "should not be given brain biopsies for fear of revealing they have no brains," Thomason reported.

In the novel, only the scientist is competent and the "only hope for salvation from the plague is a well-financed, well-traveled scientist, such as Hatfill himself."

CHAPTER 16

Resignation and Redemption

September 2002
FORT DETRICK, MARYLAND

Decked out in a starched white shirt and Armani suit, John Ezzell had the look of a changed man. The worry lines that creased his face during the winter had lifted; his scruffy beard and mustache had been carefully groomed. He was dressed for company. An army assistant secretary was visiting USAMRIID that afternoon for the official tour—capped, as usual, by peeking inside the hospital's "Slammer" and the "hot" BSL4 labs.

Even his shoe box of an office looked like the beneficiary of a toxic cleanup. The room still reflected his offbeat personality: Harley-Davidson posters adorned the walls, and the sign on the door welcomed visitors to HOTEL CALIFORNIA: YOU CAN CHECK OUT ANYTIME YOU LIKE, BUT YOU CAN NEVER LEAVE. But the stacks of paperwork that had mounted during the anthrax attacks had been gathered up, moved out, and organized elsewhere into dozens of stackable file cabinets. He could see his desk again.

The makeover seemed appropriate to Ezzell's newly enhanced status. He had just been named senior scientist to USAMRIID's new commander, Colonel Henchal, his former boss in the diagnostic systems division. One focus of the new job was to help Ridge's homeland security office create a state-of-the-art forensics laboratory, as part of a National Biological Analysis Center to be housed at Fort Detrick. Despite its lofty-sounding title, Ezzell balked at calling the new job a promotion—he joked that he was still looking for tangible proof of it in his paycheck. But it had moved him out of the daily maelstrom into a more thoughtful planning role. At the very least, it was a token acknowledgment of Ezzell's contributions to the FBI's anthrax case. More broadly, it showed the federal government's recognition that the evidentiary challenges posed by the bioterrorist attacks of 2001 were likely to be repeated.

On this clear September morning, the mood of the entire special pathogens laboratory staff felt lighter, as if an enormous burden had been lifted. Laughter wafted from a small conference room where staffers pulled sandwiches from lunch boxes. Months earlier, the room had served as a frantic command center, hooked up to others throughout the federal government.

Yet even as they had worked around the clock processing samples for the FBI and other agencies, law enforcement had considered them as possible suspects. The most senior and trusted among the special pathogens researchers, Ezzell included, had to pass FBI polygraphs and answer questions about past associations, and about whom and what they might have seen around USAMRIID's labs.

Nearly a year into the case, it was widely suspected that the anthrax used in fall 2001 must have come from within USAMRIID. Having pledged confidentiality to the FBI and the U.S. Attorney's Office, Ezzell would not discuss details of the material he found inside the deadly letters, though reporters constantly pressed him. He would only insist that there is "no indication yet that this powder or this strain that was used—that it came directly from here."[179]

At times, Ezzell felt that the scrutiny applied to his colleagues and the speculation being spread by the news media bordered on persecution and could ultimately be detrimental to the criminal case. Dedicated researchers found themselves portrayed as mad scientists, sneaking in at night for secret experiments or spiriting away spores in coat pockets. Those types of accusations had never tarnished the USAMRIID that Ezzell had known for twenty years. That was not to say, of course, that someone could not have found a way to use

USAMRIID in a devious bioterrorism plot, he realized. But how could USAMRIID stop a determined murderer?

Ezzell empathized with the young USAMRIID scientists who, at the beginning of their careers, found themselves immersed in a pressure cooker, trying to meet job demands and still find time for spouses and children. "I thought people would be right on our doorsteps, patting our backs. But we've been put in a position where we've been accused of not maintaining our cultures in a secured manner, and yet we had some of the best security in the country!" he said.

"You've worked so hard, and what have you gotten out of this? You get accused, and it's really disheartening. We're scientists. We know there's a certain amount of necessity for all of this, but at the same time we're scientists—very dedicated, very loyal, very patriotic. And as hard as we've worked, to now be subjected to these kinds of observations is demeaning."

The anthrax investigation took a steep toll on USAMRIID's labs, causing soured home lives, stress-related illnesses, early retirements.

Once, during the chaos of the spring, Ezzell had become so obviously worn down by weeks without a break that his boss ordered him to take a few days off. Go home, eat something besides junk food, get some exercise. To enforce the instruction, he stripped Ezzell of his security badge, without which he couldn't enter USAMRIID's building.

Ezzell considered a trip home to Concord, North Carolina, to catch up with family. It had been a while since he had seen his cousin Larry.

His cousin had been intrigued by the FBI's profile of the likely anthrax killer: a fiftyish white male loner with keen scientific know-how and access to anthrax spores. He jokingly asked John if it might be him. "I told John, when he comes down here I'm turning him in, and I'm gonna get the reward," Larry Ezzell said, laughing.[180]

The trip to North Carolina never materialized. Each time Ezzell prepared to leave, another urgent call would come from USAMRIID. A colleague would report the discovery of yet another possible anthrax letter bound for Fort Detrick and would summon him wearily back to the lab.

Jeffrey Koplan leaned back in a wing chair at the University Club in downtown Washington, recounting the events that had led him to his new position in life—executive vice president for academic health affairs at Emory University's Woodruff Health Sciences Center. Visiting the Capitol on business, he had just begun to adjust to academia's slower pace. From his impressive new office, he could still look out over leafy Clifton Road as he had from his old quarters in the CDC director's office. He looked rested, calm, and slightly bored.

He had made the decision to leave the CDC on a January flight back from China. In Beijing, he had been part of a ceremony honoring the twenty-year partnership between American and Chinese health officials, a project Koplan had helped launch after President Nixon reopened U.S.–China relations. In tribute, the Chinese had renamed their Academy of Preventive Medicine as the China Center for Disease Control and Prevention—the "China CDC," as it would become known. During a ceremony, Minister of Health Zhang Wenkang had appointed Koplan as the agency's first senior adviser.

The event gratified Koplan, who had devoted untold time to helping the Chinese upgrade their abysmal health system, spread thin over a poverty-racked countryside of remote villages and provinces. Global outreach had been one of Koplan's priorities as CDC director and was one of the reasons he decided to stay on for an extra year when the Clinton administration departed. He enjoyed strong Republican backing on Capitol Hill, and the incoming Bush administration had shown no desire to remove him from the post.

When the anthrax attacks began, Koplan shelved all thoughts of leaving the CDC, determined to see the agency through a national emergency.

He thought, *This is what I've spent thirty years training for, and I'm in the right place with a great group of people to work on a major national problem, and I'm lucky to be here to be able to work on it. No one would wish this on anybody, but if it's gonna happen, better to be part of it and try to solve it than be on the sidelines.*

With the crisis finally under control by January 2002, Koplan used the long trip from Beijing to Atlanta to contemplate the reality of his working life. His friend Dr. Brachman had once described as a blessing the distance between CDC's offices in Atlanta and HHS headquarters in Washington. But Koplan had found that the geographic buffer no longer shielded him from the pressures of the political arena. Under HHS Secretary Thompson, he found his job pressures increasingly hard to tolerate. He had come to believe that the CDC would be better served by someone more attuned to Thompson's management style.

Koplan had not felt slighted when Thompson surrounded himself on public platforms with what seemed like every other federal health official but the CDC director. There was too much to do and plenty of credit to go around, he thought.

His problems arose when his duties outstripped his authority to carry them out. Since the beginning of the anthrax attacks, Thompson and his political appointees had demanded input into CDC decisions, both major and minor, and into the flow of information to the public.

"The new team had their own management style, and it was different from the one I was comfortable with, and that created regular tensions and stresses," Koplan said. "I think they had a view of what the CDC was. They would say things along the lines of '[He's] not a team player,' and I'd try to combat that. Even though we were there in Atlanta, we thought of ourselves as very much part of a national health effort."

Under Thompson, HHS's strategy at the start of the crisis had been centralized, airtight information control. Political types at HHS, rather than the CDC doctors on the front lines, responded to press questions. "It's the opposite of our normal approach," Koplan said, "which is [to] push the press relationship to the closest scientist who can do it, as long as that person can walk and chew gum at the same time."

Once Thompson's office lifted its restrictions, the CDC Web site offered a constant stream of health advisories and guidelines, with Koplan and other doctors answering questions in regular press telebriefings. The belated effort won praise, but the impression had already been created that health officials were dodging questions, or worse yet, covering something up.

Microbiologist Salyers and other scientists were disturbed that politics, and what appeared to be a CDC "gag order" only escalated the health crisis. [181] A Harvard School of Public Health survey conducted in early November showed that no national official was seen by a majority of the public as a reliable source of information during a bioterror event. The CDC director ranked the highest among federal officials, but citizens were more inclined to trust local health officials.[182]

Koplan had fought attempts to make the CDC the scapegoat for what he continued to view as an unavoidable public health tragedy, with lessons that could be learned only through hard experience. "In every single instance when something turned sour, when the anthrax hit the fan, the finger was always pointed at the CDC," he said.[183] Yet as Dr. Mohammad N. Akhter, the president of the American Public Health Association, put it, the CDC's anthrax challenge was like "trying to build a plane while flying."[184]

The postal service, in Koplan's opinion, had stopped just short of blaming the CDC for the deaths of postal workers, as if to deflect attention from their own role in the decision making. Postal officials grew frustrated that the CDC could not give "clear-cut courses of action," Koplan said, yet decisions about whether postal facilities should close clearly rested in the hands of postal authorities, not doctors.[185]

From the CDC's perspective, he said, the surest way to protect lives would have been to tell postal managers, "If this is too complicated, then shut everything down! That would have been the easiest course of action for us, but never once did we threaten that or say that out of anger. Never once."

Like John Ezzell, Koplan felt the harsh sting of public criticism. On October 24, the Senate Appropriations Committee's Public Health Subcommittee called Koplan to testify about the CDC miscalculations at Brentwood. Koplan, his foot still mending, arrived to find a wheelchair and a top committee aide at the hearing-room doorway. Waiting to be wheeled inside, the aide prepared him to face a firing squad.

She delivered "the old don't-take-it-personally routine," Koplan recalled. "So I just sucked it up and thought, 'Well, this is just one of those days.'"

Chairman Tom Harkin (D-Iowa) led the pack, holding up that day's *Washington Post* story on Brentwood and asking Koplan to justify the contrasts between the attentive response on Capitol Hill and the slow-footed Brentwood reaction.

Koplan, bristling, fought to keep from being defensive. He had faced congressional inquisitions before and knew it was best simply to admit mistakes and promise that they would not be repeated.

Yet, as he said, "There's nothing more disturbing to us in public health and to me personally than the idea that there was a dual standard for white, well-educated Senate staff versus for African-American, blue-collar postal workers."

Koplan knew that the series of racially charged events surrounding the Brentwood deaths fit a stereotypical model straight out of Hollywood. But he also knew that "if you gave any person in public health a choice and said, 'You've only got an hour to do some important health work, and your choice is to do it in the Senate or with this group of blue-collar workers,' there's not a public health person that wouldn't go to the blue-collar workers. But these things don't unfold in ways that you can manage."

Koplan felt defenseless. He wished that a senator would just accuse him of stupidity, of misinterpreting clear data. "Then I'd say, 'Well, we were dumb. We didn't do the job. That's terrible, that's distressing, our professional stature is challenged. But you're right,'" he said.

Instead, his agency's "value system was being maligned," an unbearable assault. He struggled to get through the hearing, doing his best to describe "what we did, why we did it, and be both sincere and honest about it."

Koplan left the hearing room with reporters buzzing around him. One questioner asked about the public health service's involvement in the Tuskegee experiments, suggesting that the CDC had a racist past. "At which point, I just headed out in the wheelchair and just left," he said.

The next day, Koplan called Harkin, whom he had regarded as a friend and a consistent supporter of the CDC. When the senator came to the phone, he began revisiting the Brentwood discussion, but the CDC director cut him short.

"Let me explain something to you," Koplan told Harkin. "We are in the midst of an attack, and we are your defensive force. We may screw up, and we may do things wrong, and in retrospect, people may look back and say, 'They should have done this then' or 'dug that trench there. . . .' But at this moment, we've got everybody working full steam on this, twenty-four hours a day, with nothing but the intent of stopping this and solving it. We need your support."

Harkin listened quietly, then said, "You're right."

The next day, Koplan was heartened when he heard Harkin on television, echoing his argument.

Almost everyone who knew Koplan thought that he had found his calling running the CDC and that he would never leave without being pushed. He loved the job and viewed its hurdles as challenges, its political land mines as conquerable obstacles. But the Emory job proved irresistible.

In February, Koplan flew to Washington to inform Thompson that he would take the Emory post and step down the next month. Thompson said he was saddened by Koplan's decision and considered him "a great scientist" who had done a laudable job.[186] Thompson appointed an interim management team with an overseer at HHS headquarters, then took his time filling the job. In July, he chose Julie L. Gerberding, the forty-six-year-old South Dakota native who had served as acting principal deputy director at CDC. "The events of last fall made it clear to all of us that this cannot be a time of business as usual," Gerberding said upon accepting the position.[187]

In a press release announcing his resignation, Koplan called the CDC director's job "the best job in public health." At the top of his list of accomplishments, he cited "responding swiftly and effectively to the nation's first major bioterrorism event."

The conclusions reached by the American public ran the gamut. Some faulted the CDC for not acting quickly enough to administer antibiotics; others questioned the medical judgment of dosing thousands who never were truly at risk.

A sympathetic review came from anthrax specialist Philip Brachman, who was asked by a Senate committee to evaluate the CDC response. A major problem, he said, was that the CDC "leadership was subverted." With a competent and trained staff, ready to travel at a moment's notice, the agency "must be given the authority to operate as it always has in times of emergency. To put dampers on its actions can only lead to problems."[188]

Koplan left the CDC confident that the agency would deal more effectively with the next bioterrorist event, partly through his efforts. He took pride in Popovic's anthrax laboratory, the CDC-trained state lab directors, the response teams that had rushed medicine and doctors into the field. His successor, Gerberding, vowed to address other fundamental issues—to make the CDC culture more responsive and adaptable.

During Koplan's last weeks in office, Harkin and Senator Arlen Specter (R.-Pa.), the ranking Republican on the Senate Labor, Education and Health and Human Services Committee, sponsored a resolution honoring Koplan's service. In a floor statement, Specter lauded Koplan for "working around the clock to prevent people exposed to anthrax from developing the disease."[189]

House and Senate leaders honored the doctor at an elegant reception. On Capitol Hill, where the CDC had performed at its best, Koplan was the man of the hour.

In 2005, John Ezzell becomes eligible for federal retirement, but he intends to stay in the job until at least 2008. That should give him time to see the new forensic laboratory come into being.

His greatest hope is that the FBI will catch the anthrax killer. The criminal, he said, has tarnished the purity of science; researchers at USAMRIID and other federal agencies are now obliged to work under surveillance cameras. Lab security has become a hot topic, as all researchers feel the impact of a criminal's actions.

The prospect of helping to catch a killer motivated Ezzell through the most stressful days. He anticipates taking the witness stand to testify about his careful custody of the evidence, about the opening of the letters and the wisp of powder dispersing in the air. He wants to know that the public is safe bringing in the mail without sterilizing it in the kitchen oven.

The attacks have changed his way of looking at the world. One recent summer morning, he got up before dawn, grabbed a life jacket, and paddled his kayak along the Potomac River.

"I got to the middle of the river, and I flipped," he said. He struggled, trying to right himself, then laughed imagining the newspaper headlines and the conspiracy theories they might inspire: ANTHRAX EXPERT DROWNS IN RIVER.

He paddled back to shore and dragged the kayak back into his yard, near the peaceful lily ponds and gardens.

The next day he was back at work in Fort Detrick, building defenses against deadly pathogens and realizing, as his cousin Larry put it, that "time marches on."

CHAPTER 17

Home Again

October 19, 2002
STAFFORD, VIRGINIA

On the first anniversary of Leroy Richmond's anthrax diagnosis, his wife, Susan, laid down the law. She had seen quite enough of the nosy reporters tromping across her spotless carpets, the television crews blocking her way as she pulled in and out of the garage, honking her car horn. Inside, the phone jangled, and it would almost always be another imposition on Rich's time—a press interview while he was trying to put Quentin to bed, a doctor's office confirming an appointment, the paper-pushers from the U.S. Department of Labor checking up on his workmen's compensation. He appreciated the attention, showing patience with the most trivial questions. Meanwhile, she stewed.

As October 19 approached, he had jotted down more interviews and meetings, but when he looked at his calendar again a few days later, everything was crossed out.

"You tell them you're spending the day with your family!" she bellowed as he entertained a request over the telephone receiver. "With your *wife* and your *son!* You tell them that! You hear me?"[190]

"Yes, dear," Rich replied softly, doing his best to humor her. In reality, he had hoped to fill up the day, to keep memories from overpowering him. He could still find himself overcome with panic and silent rage over the deaths of his friends, the decision to leave Brentwood open when its danger should have been obvious.

After his weeks in the hospital, Rich had emerged as a man with a mission, determined to inform the public about his anthrax battle and to bring recognition to the postal workers who contracted the disease. He had just signed up for a medical study of the survivors, to be conducted with funding from NIH. Rich could not fathom why it had taken government doctors so long to realize the survivors might have something to teach.

Rich had a plan to compensate for the government's indifference to the anthrax victims and survivors by establishing a charity, the Anthrax Relief and Scholarship Fund. "This is for all anthrax victims, also the people who were exposed in New Jersey and New York," he said. The victims had formed their own support network, talking frequently to share medical updates, nightmares, and fears.

Rich could not imagine returning anytime soon to the decontaminated Brentwood postal facility. Nor could Susan, who never wanted to see the place again. Rich still saw a psychiatrist once a month to manage depression and panic attacks. His short-term memory was gone, a common complaint among the few inhalation survivors. He would think back on the intricate zip-code maps he used to know by heart and realize how much he had lost.

There were signs of progress, though. The appointments with medical specialists, which used to cover the family calendar in a wash of blue ink, were down to two or three a month with either the lung specialist, the heart specialist, or the infectious disease specialist. The only daily medication he took was a standard antidepressant.

With his recaptured energy, he traveled and made public appearances. Two days after the family celebration of his recovery, he would honor Morris and Curseen at a memorial service at Washington's National Shrine. In mid-August, he flew to Minneapolis, Minnesota, to speak at a convention of the American Postal Workers Union. He carried a special apron, made by Susan, emblazoned with his motto, POSTAL WORKERS ARE HEROES TOO.

At the gathering, he bonded with the widows of Curseen and Morris. The Morris family's negligence suit against Kaiser

Permanente had been settled out of court for an undisclosed sum after lawyers made known their intention to emphasize the issue of racially disparate treatment if the case came to trial.

Rich also met several African-American congressmen, who invited him to a 9/11 victims' tribute held by the Congressional Black Caucus where Morris and Curseen would be honored along with three African-American schoolchildren who were aboard the hijacked Flight 77. At the ceremony, he met District of Columbia Mayor Anthony Williams, the official who had gone on the air the night of October 19, 2001, to announce that an unnamed Brentwood worker had been admitted to the hospital in critical condition. Rich had watched the broadcast from his hospital bed, wondering how the news about him had leaked out so quickly.

Rich went over to shake the mayor's hand, expecting a warm greeting. Instead, he saw a blank look and realized, "He had no idea who I was. But that's politics," Rich said.

Postmaster General Potter, also in attendance, found Rich after the ceremony to shake his hand and express his concern. The men chatted cordially until one of Potter's aides alerted him that he was running late for a downstairs reception. Potter took Rich along with him, where they hobnobbed with caucus leaders from Capitol Hill.

Driving back and forth to Washington, Rich still pulled out his rosary beads and used the time to pray, just as he did a year ago on his commute to Brentwood. World peace still seemed elusive. He fretted about Bush's plan to attack Iraq; he had a nephew in the army, recently married, and he feared for his safety.

Sometimes, Rich pondered the identity of the anonymous anthrax killer. On television, he had watched Steven Hatfill's dramatic denials. He had read news reports suggesting that the FBI, after a year of work by hundreds of agents across the country, was no closer to solving the crimes.

Rich had his own theory about the perpetrator. He believed that whoever had concocted the weaponized anthrax that found its way into his lungs most likely had not survived the attacks. The material was simply too dangerous to be handled without harm.

One thing seemed certain. The perpetrator had not set out to specifically target Mo Morris or Joe Curseen or to make Leroy

Richmond or Norma Wallace or Ernesto Blanco deathly ill. Cloaked in anonymity, the killer almost certainly did not know Bob Stevens or Kathy Nguyen or Ottilie Lundgren. The chaos created when fine anthrax went through high-speed postal machines probably came as a complete surprise.

The killer succeeded, however, in delivering a bloodcurdling message. Bioterror was not a wild fantasy, but as real as the next trip to the mailbox. It could strike anywhere, anytime. Open an envelope packed with powder, and if you were lucky, the scientists in John Ezzell's laboratory would find that it contained baby powder or lumpy tobacco. If not, it just might be a high-grade weaponized anthrax capable of wiping out a U.S. city.

In some sense, the Chicken Littles were right: the sky *was* falling. The only sure way to survive was to listen to the prophets, to infuse their mission with the money and respect so long denied them.

In the months after the anthrax attacks, the Bush administration rethought its early plans to downscale bioterrorism preparedness spending and instead recommended a gigantic increase, more than doubling the annual expenditure to $1.5 billion. Officials at the CDC, USAMRIID, and a host of other agencies worked frenetically to upgrade their laboratories and training programs, trying to narrow the discomfiting odds that the bumbling responses to the 2001 anthrax letters would happen again.

Rich tried not to dwell on imponderables. Instead, he savored the fact that he could walk unaided to the corner school bus stop to meet Quentin. Sometimes, he even found the strength to kick a soccer ball around the front yard.

He could open the front door of their Stafford house and see the comforting family portraits, the dining-room table set as if for company. He could peek from behind the curtains and see the mailman working his route along their quiet suburban cul-de-sac, stuffing the boxes with letters and junk mail and bundles of Christmas catalogs.

He could draw a deep breath and almost imagine life returning to normal, as it was before the whoosh of white powder changed everything.

NOTES

This book is based largely on interviews conducted between April 2002 and October 2002 by the author and her researchers, Davene Grosfeld and Maryanne Warrick. Major written sources included publications of the U.S. Centers for Disease Control, the World Health Organization, the U.S. Postal Service, the U.S. Army Medical Research Institute for Infectious Diseases, and other government agencies; medical and scientific journals; congressional hearing records; court files; contemporaneous news reports and government documents supplied to the author.

1 The author and researcher D. Grosfeld conducted numerous interviews of Leroy Richmond from May to October 2002, both at his home in Stafford, Virginia, and by telephone. The portrait in this chapter is based on material from all of them.

2 Richmond, Susan, interview by D. Grosfeld, Stafford, VA; June 19, 2002.

3 Richmond, Leroy, interview by author, Stafford, VA; April 23, 2002.

4 Centers for Disease Control, "Human Anthrax Associated with an Epizootic Among Livestock," *Morbidity and Mortality Weekly Report,* Vol. 50, No. 32, August 17, 2001.

5 Centers for Disease Control, "Evaluation of Bacillus Anthracis Contamination Inside the Brentwood Mail Processing and Distribution Center—District of Columbia October 2001," *Morbidity and Mortality Weekly Report,* Vol. 50, No. 50, December 21, 2001.

6 Hugh-Jones, M. E., "Global Report, 2000," 4th International

Conference on Anthrax, Program and Abstracts Book, June 10–13, 2001, Annapolis, MD, p. 13.

7 Marston, C., et al., "Training the Trainers: Building Bioterrorism Response Capacity to Rapidly Isolate and Identify Bacillus Anthracis," 4th International Conference on Anthrax, Program and Abstracts Book, June 10–13, 2001, Annapolis, MD, p. 21.

8 Turnbull, Peter C. B., interview by author, Silver Spring, MD; April 28, 2002.

9 Ezzell, Larry, telephone interview by D. Grosfeld; October 8, 2002.

10 Ezzell, John W., interview by author, Frederick, MD; June 26, 2002.

11 Friedlander, Arthur, telephone interview by author; July 12, 2002.

12 Franz, David, interview by author, Salt Lake City, UT; May 20, 2002.

13 Siller, Bobby L., telephone interview by author; May 13, 2002.

14 Harris, Larry Wayne, telephone interview by author; May 2, 2002.

15 Batt, T., "Anthrax Incident Helps FBI," *Las Vegas Review Journal,* April 23, 1998.

16 Patrick, William C., III, interview by author, Frederick, MD; May 7, 2002.

17 Covert, Norman M., *Cutting Edge: A History of Fort Detrick, Maryland,* 4th Edition, October 2000. Originally published May 1993.

18 Jemski, Joseph, interview by M. Warrick, Frederick, MD; July 16, 2002.

19 Sidell, Frederick, E. T. Takafuji, and David R. Franz, *Textbook of Military Medicine, Medical Aspects of Chemical and Biological Warfare,* Washington, D.C., Borden Institute, Walter Reed Medical Center, Office of the Surgeon General, 1997, Chapter 19 (Cheryl, Franz, D. Parrott, and Ernest T. Takafuji, "The U.S. Biological Warfare and Biological Defense Programs").

20 Patrick interview, 5/7/02.

21 Boyles, Charles, telephone interview by M. Warrick; May 14, 2002.

22 U.S. Army, "Report on Demilitarization of Fort Detrick," Appendix F: Fact Sheet on Three Laboratory Occupational Deaths, 1951, 1958 and 1959; 1977.

23 Patrick interview, 5/7/02.

24 "Report on Demilitarization of Fort Detrick."

25 Patrick interview, 5/7/02.

26 Torres, Arthur R., telephone interview by M. Warrick; May 29, 2002.

27 Patrick interview, 5/7/02.

28 Torres interview, 5/29/02.

29 Tsonas, Christos, telephone interview by author; July 5, 2002. Tsonas referred to William J. Broad and David Johnston, "Report Linking Anthrax and Hijackers Is Investigated," *New York Times,* March 23,

2002, as the best account of the emergency-room encounter.

30 Breed, A., "Third Case of Anthrax Exposure in Florida; Federal Authorities Open Criminal Investigation," Associated Press, October 11, 2001.

31 Tyler, Patrick E., with John Tagliabue, "Czechs Confirm Iraqi Agent Met with Terror Ringleader," *New York Times,* October 27, 2001.

32 Lester, James, telephone interview by author, July 2002.

33 *United States of America* v. *Zacarias Moussaoui,* December 2001 term, Eastern District of Virginia, Alexandria.

34 CNN.com, "New Bin Laden Tape Surfaces," April 16, 2002. Available at http://www.cnn.com/2002/WORLD/meast/04/15/terror.tape/index.html.

35 O'Toole, Tara, telephone interview by author; May 3, 2002.

36 Broad, William J., and David Johnston, "Report Linking Anthrax and Hijackers Is Investigated," *New York Times,* March 23, 2002. Available at http://www.ph.ucla.edu/epi/bioter/anthraxhijackerslink.html.

37 Gentile, Don, telephone interview by author; July 1, 2002.

38 Brachman, Philip, interview by author, Atlanta, GA; June 11, 2002.

39 Ostroff, Stephen M., interview by author, Atlanta, GA; June 10, 2002.

40 Brachman interview, 6/11/02.

41 Perkins, Bradley, interview by author, Atlanta, GA; June 11, 2002.

42 Jerrigan, J., et al., "Bioterrorism-Related Inhalational Anthrax: The First Ten Cases Reported in the United States," *Emerging Infectious Diseases,* Vol. 7, No. 6 (November-December 2001), pp. 933–944.

43 Malecki, Jean Marie, interview by author, Palm Beach, FL; June 17, 2002.

44 Malecki, Jean Marie, Epidemiologic study by, provided to author.

45 Perkins interview, 6/11/02.

46 Zaki, Sherif interview by author, Atlanta, GA; June 10, 2002.

47 Popovic, Tanja, interview by author, Atlanta, GA; June 10, 2002.

48 Ibid.

49 Blanco, Ernesto, interview by author, Boca Raton, FL; June 18, 2002.

50 Bush, George W., Remarks by the President in Photo Op with Speaker Hastert, Leader Daschle, Minority Leader Lott, and Minority Leader Gephardt, The Oval Office, Washington, D.C.; October 2, 2001.

51 Koplan, Jeffrey, interview by author, Atlanta, GA; June 11, 2002.

52 "Are We Safe?," *60 Minutes,* CBS News, September 30, 2001. Available http://www.cbsnews.com/stories/2001/09/30/60minutes/printable313043.shtml.

53 Frist, Bill, *NewsHour with Jim Lehrer,* PBS, October 3, 2001.

54 U.S. Department of Health and Human Services, "Secretary Thompson Testifies on Bioterrorism Preparedness: Statement Before the Senate Appropriations Subcommittee on Labor, Health and Human Services, Education and Related Agencies," Press Release, October 3, 2001. Available at http://www.hhs.gov/asl/testify/t011003.html.

55 "Ready or Not: Report by Kwame Holman on Congressional Hearings Regarding Bioterrorism," *Online NewsHour with Jim Lehrer,* October 3, 2001. Available at http://www.pbs.org/newshour/bb/terrorism/july-dec01/kwame_10-3.html.

56 Frist, Bill, Testimony Before the Senate Appropriations Committee Subcommittee on Labor, Education, Health and Human Services, October 3, 2001.

57 Frist, Bill, interview by author, Washington, D.C.; May 17, 2002.

58 Centers for Disease Control, *Emerging Infectious Diseases,* Vol. 4, No. 3 (July–September 1998).

59 Thompson, Tommy, telephone interview with author; November 2002.

60 Fleischer, Ari, White House Press Briefing, October 4, 2001, Washington, D.C., http://www.whitehouse.gov/news/releases/2001/10/20011004-12.html.

61 Bresnitz, Eddy, interview by D. Grosfeld, Trenton, NJ; June 4, 2002.

62 "Florida Man Hospitalized for Anthrax," Associated Press, October 4, 2001.

63 Koplan interview, 6/11/02.

64 Garrett, Laurie, "Clash of Agencies Hampered Inquiry into Anthrax Mystery," *Newsday,* July 23, 2002.

65 Cohen, Mitchell, telephone interview by author; August 12, 2002.

66 Gerberding, Julie, telephone interview by author; June 11, 2002.

67 News Conference with Attorney General John Ashcroft, Department of Justice, Washington, D.C., October 8, 2001.

68 Administration Officials Discuss Anthrax Situation in U.S.; Ridge, Ashcroft, Mueller, Others Briefed at Old EOB, October 18, 2001. Available at http://usinfo.state.gov/topical/global/hiv/01101806.html.

69 Salyers, Abigail, interview by author, Salt Lake City, UT; May 21, 2002.

70 Huden, Johanna, telephone interview by D. Grosfeld; May 30, 2002.

71 Zaki, Sherif, interview by author, Atlanta, GA; June 10, 2002.

72 Koplan, Jeffrey, telephone interview by author; August 9, 2002.

73 "One Anthrax Case in NYC; Tests in Nevada Inconclusive," October 12, 2001. Available at http://www.cnn.com/2001/HEALTH/conditions/10/12/nyc.anthrax/.

74 Ostroff interview, 6/10/02.
75 Maraynes, Allan, telephone interview by D. Grosfeld; July 29, 2002.
76 "Osama's Enabler in Congress," Human Events, *The National Conservative Weekly,* December 3, 2001, p. 1.
77 Cheney, Richard, The Vice President Appears on *NewsHour with Jim Lehrer,* PBS Network, Washington, D.C., October 12, 2001.
78 Bush, George W., Remarks by the President in Photo Op with Italian Prime Minister Silvio Berlusconi, White House Colonnade, Washington, D.C., October 15, 2001.
79 Bresnitz interview, 6/4/02.
80 DiFerdinando, George, interview by D. Grosfeld, Trenton, NJ; June 4, 2002.
81 Bresnitz interview, 6/4/02.
82 DiFerdinando interview, 6/4/02.
83 Bresnitz interview, 6/4/02.
84 Parker, Major General John S., commanding general, U.S. Army Medical Research and Matériel Command, Testimony Before the Committee on Governmental Affairs and the Subcommittee on International Security, Proliferation and Federal Service, October 31, 2001.
85 Smith, James R. E., telephone interview with M. Warrick; July 18, 2002.
86 News Conference with FBI Director Mueller and Attorney General Ashcroft, Washington, D.C., October 16, 2001.
87 National Center for the Analysis of Violent Crime, Federal Bureau of Investigation, Linguistic/Behavioral Analysis of Anthrax Letters, Critical Incident Response Group, November 9, 2001.
88 Ezzell, John W., "Procedure for Killing Bacillus Anthracis Spores in Mail." Posted at www.gbgmumc.org/middletown.
89 Potter, John, interview by author, Washington, D.C.; August 26, 2002.
90 U.S. Department of State, Administration Officials Discuss Anthrax Situation in U.S., White House Press Briefing, October 18, 2001. Available at http://usinfo.state.gov/topical/global/hiv/01101806.htm.
91 Ibid.
92 Ibid.
93 United States Postal Service, "U.S. Postal Inspection Service, FBI Offer Reward of up to $1 Million for Information Leading to Arrest of Anthrax Mailer," Press Release, October 18, 2001.
94 Ibid.

95 Koplan interview, 8/9/02.

96 Daschle, Thomas, "Remarks by Senate Majority Leader Tom Daschle Concerning the Anthrax Attack on Capitol Hill Delivered on the Senate Floor," October 18, 2001.

97 Lancaster, John, and Helen Dewar. "N.J. Mail Carrier, CBS Employee Have Anthrax," *Washington Post,* October 19, 2001, A12, Final Edition.

98 Koplan interview, 8/9/02.

99 Centers for Disease Control, "Bioterrorism Alleging Use of Anthrax and Interim Guidelines for Management—United States, 1998," *Morbidity and Mortality Weekly Report,* Vol. 48, No. 4, February 5, 1999, page 4.

100 Kournikakis, B., S. Armour, C. Boulet, M. Spence, B. Parsons, "Risk Assessment of Anthrax Threat Letters," Report No. DRES-TR-2001-048, Defense Research Establishment Suffield, September 2001.

101 Koplan interview, 8/9/02.

102 Potter interview, 8/26/02.

103 DiFerdinando interview, 6/4/02.

104 Mayer, Thom, telephone interview by D. Grosfeld; May 2, 2002.

105 Murphy, Cecele, telephone interview by D. Grosfeld; May 7, 2002.

106 Susan Richmond interview, 6/19/02.

107 Richmond, Leroy, telephone interview by D. Grosfeld; March 17, 2002.

108 Leroy Richmond interview, 4/23/02.

109 Susan Richmond interview, 6/19/02.

110 Bresnitz, Dr. Eddy A., and Dr. George T. DiFerdinando Jr., "Lessons from the Anthrax Attacks of 2001–The New Jersey Experience," manuscript to be published in 2003.

111 "A Postal Worker's Correct Fear; 911 Call Shows D.C. Postal Worker Suspected Anthrax Infection Before His Death," November 7, 2001. Available at http://abcnews.go.com/sections/us/DailyNews/ANTHRAX_Morris911.html.

112 Becker, Elizabeth and Robin Toner, "Postal Worker's Illness Set Off No Alarms," *New York Times,* October 24, 2001.

113 Susan Richmond interview, 6/19/02.

114 Potter interview, 08/26/02.

115 Ballard, Tanya, "Families, Colleagues Grieve for Postal Workers Who Died from Anthrax," November 13, 2001. At http://govexec.com/.

116 Ibid.

117 Willhite, Deborah, telephone interview by author; September 5, 2002.

118 Ibid.

119 Koplan, Jeffrey, interview by author, Washington, D.C.; September 4, 2002.
120 White House Press Briefing, Washington, D.C.; October 22, 2001.
121 Khabbaz, Rima, telephone interview by author; June 28, 2002.
122 Perkins, Bradley, telephone interview by author; August 12, 2002.
123 Khabbaz interview, 6/28/02.
124 Koplan interview, 8/9/02.
125 Khabbaz interview, 6/28/02.
126 Ostroff, Stephen M., telephone interview by author; August 2, 2002.
127 Centers for Disease Control, "Update on Anthrax Investigations with Drs. Jeffrey Koplan and Julie Gerberding," November 7, 2001. Available at http://www.cdc.gov/od/oc/media/transcripts/t011107.html.
128 Ostroff interview, 8/2/02.
129 New York City Department of Health, Office of Public Affairs, "Update on Anthrax Situation in New York City," Press Release, November 9, 2001.
130 Ostroff interview, 8/2/02.
131 Mark Uehling, "Acting on Anthrax: What One Lab Learned," *CAP Today*, The College of American Pathologists, February 2002.
132 Crowther, Peg, telephone interview with M. Warrick; July 23, 2002.
133 Albee, Luke, and Clara Kircher, interview with M. Warrick, Washington, D.C.; June 14, 2002.
134 Albee, Luke, Interoffice Memo to Senator Patrick Leahy, October 12, 2001.
135 Leahy, Patrick, interview by author, Washington, D.C.; July 25, 2002.
136 Eggan, Dan, and Susan Schmidt, "Fourth Anthrax Letter Discovered by FBI," *Washington Post,* November 17, 2001, p. A1.
137 Leahy interview, 7/25/02.
138 Leahy, Patrick, "Comment . . . on the FBI's Discovery of a Suspicious Letter . . . ," November 16, 2001. Available at http://leahy.senate.gov/press/200111/111601a.html.
139 Senate Judiciary Committee, Subcommittee on Technology, Terrorism and Government Information, Hearing Transcript, November 6, 2001.
140 National Center for the Analysis of Violent Crime, Federal Bureau of Investigation, Linguistic/Behavioral Analysis of Anthrax Letters, Critical Incident Response Group, Amerithrax Press Briefing, November 9, 2002.
141 Ibid.
142 Leahy interview, 7/25/02.
143 Ezzell, John W., interview by author, Frederick, MD; June 26, 2002.

144 McKelvey, Gerald, telephone interview by author; May 6, 2002.
145 Friedlander, Arthur, telephone interview with M. Warrick; August 8, 2002.
146 Jerrigan et al., "Bioterrorism-Related Inhalational Anthrax."
147 Kramer, Robert, interview with M. Warrick; September 5, 2002.
148 Martin, Tim, "Pentagon Assessing BioPort," *Lansing State Journal,* August 24, 2001. Available at http://www.lsj.com/news/bioport/010824_bioport_1a-6a.html.
149 Dunn, Barbara, "The Anthrax Vaccine Immunization Program—What Have We Learned?," Testimony Before the Committee on Government Reform, House of Representatives, October 3 and 11, 2000. Serial No. 106-249.
150 Brennan, Phil, "Sole Source of Anthrax Vaccine Isn't a Source at All," Newsmax.com, October 8, 2001. Available at http://www.newsmax.com/archives/articles/2001/10/6/01001.shtml.
151 Martin, Tim, "FDA Trying to Assist BioPort," *Lansing State Journal,* October 28, 2001. Available at http://www.lsj.com/news/bioport/011028_bioport_3a.html.
152 Evenson, A. J., and Tim Martin, "BioPort Aims Toward New Contract," *Lansing State Journal,* June 11, 2001. Available at http://www.lsj.com/news/bioport/bioport_000611a.html.
153 The Institute of Medicine, "The Anthrax Vaccine: Is It Safe? Does It Work?," March 7, 2002. Available at http://www.nap.edu/html/anthrax/index.html.
154 U.S. Department of State, "Pentagon Announces Resumption of Anthrax Vaccine Program," International Information Programs, June 29, 2002. Available at http://usinfo.state.gov/topical/global/hiv/02062903.html.
155 Kramer interview, 9/5/02.
156 "Double Exposure," Editorial, *Washington Post,* December 19, 2001, p. A38.
157 Day, Thomas G., telephone interview by author; August 26, 2002.
158 Ibid.
159 Statement of Steven J. Hatfill, August 25, 2002.
160 Kristoff, Nicholas, "The Anthrax Files," *New York Times,* August 13, 2002.
161 CNN.com, "Ashcroft: No Charges Yet in Anthrax Probe," August 22, 2002.
162 Clawson, Patrick, speaking to reporters at Hatfill press conference, August 25, 2002.

163 Hatfill submitted a biographical sketch and photo from a BSL4 laboratory to his college.
164 Résumé of Steven J. Hatfill, posted on the Internet around 1999. Available at http://www.anthraxinvestigation.com/hatfill.pdf.
165 *Newsweek,* August 12, 2002.
166 Franz, David, telephone interview by author, October 4, 2002.
167 Hatfill résumé.
168 Salyers interview, 5/21/02.
169 *Armstrong Williams Show,* "The Emerging Threat of Biological and Chemical Terrorism," 1998.
170 Potomac Institute for Policy Studies, Proceedings Report, June 16, 1998.
171 Reed, Fred, "Biological Terrorism Is a Real Threat," *Washington Times,* August 11, 1997.
172 CBSNews.com, "Ridge: Terror War a Long Haul," January 13, 2002. Available at http://www.cbsnews.com/stories/2002/01/13/attack/main324166.shtml.
173 Rumsfeld, Donald, *Fox News Sunday,* as cited in *Biohazard News,* February 26–September 10, 2001.
174 Salyers interview, 5/21/02.
175 Harp, Van, "FBI Letter to Members of the American Society of Microbiologists," January 29, 2002.
176 Rosenberg, Barbara Hatch, "Commentary: Is the FBI Dragging Its Feet?" February 5, 2002. Posted at http://www.fas.org/bwc/news/anthraxreport.htm.
177 Ibid.
178 E-mail correspondence with author, October 24, 2002.
179 Ezzell, John W. interview by author, Frederick, MD; August 29, 2002.
180 Larry Ezzell telephone interview, 10/8/02.
181 Salyers interview, 5/21/02.
182 Harvard School of Public Health, "Survey Shows Americans Not Panicking over Anthrax, but Starting to Take Steps to Protect Themselves Against Bioterrorist Attacks," Press Release, November 8, 2001.
183 Koplan interview, 9/4/02.
184 Akhter, Dr. Mohammad N., American Public Health Association, Press Release, February 22, 2002.
185 Koplan interview, 9/4/02.
186 Thompson, 11/27/02.
187 Centers for Disease Control, "Julie Gerberding, MD, MPH, Named

CDC Director and ATSDR Administrator," Press Release. Available from http://www.cdc.gov/od/oc/media/pressrel/r020703b.html.

188 Brachman, Philip S., Letter to Senator Charles Grassley, February 26, 2002.

189 "Regarding the Career of Jeffrey Koplan, M.D.," *Congressional Record,* Senate, April 26, 2002, pp. S3464–S3465.

190 Richmond, Leroy, telephone interview by D. Grosfeld, October 9, 2002.

ACKNOWLEDGMENTS

A number of people helped make this book happen. Three emerge as major characters in the narrative. Leroy "Rich" Richmond and his wife, Susan, gave unselfishly of their time to bring a human dimension to the anthrax story. John W. Ezzell, now a senior scientist at the U.S. Army Medical Research Institute of Infectious Diseases (USAMRIID), patiently demystified the science behind the greatest laboratory challenge of his career. Dr. Jeffrey P. Koplan endured questions large and small, even as he moved from director of the Centers for Disease Control to a job as vice president for Academic Health Affairs at Emory University in Atlanta. All of them wanted this story told in the hopes that it would enlighten future responses to bioterror threats.

Davene Grosfeld and Maryanne Warrick worked exceptionally hard to help me produce this book. Other help with the manuscript came from Robert W. Thompson Jr., John Bennett, and Kevin McCoy. My sons, Cory and Andrew, were dedicated assistants.

Gail Ross redefined the role of agent with her continuing interest and sound advice. At HarperCollins, my thanks for superior editing go to David Hirshey, Jeffrey Kellogg, Emily McDonald, and Nick Trautwein.

The CDC's public affairs office was responsive and professional, especially Lisa Swenarski, Llelwyn Grant, and Jennifer Morcone. The CDC's Dr. Tanja Popovic generously shared her photographs. Other significant help came from Caree Vander Linden and Chuck Dasey at USAMRIID, and the public affairs staff of the U.S. Postal Service.

At the *Washington Post*, special thanks go to executive editor Leonard Downie Jr. for giving me the time to complete this project. My colleagues Jeff Leen, Joby Warrick, Margot Williams, and Rob Thomason were helpful. For providing moral support, my gratitude goes to Ines Figueiredo, Dennis Williams, Anne Sundermann, and Lois Barrence.

INDEX